Motorcycle chassis design

Motorcycle chassis design: the theory and practice

Tony Foale and Vic Willoughby

First published in 1984 by Osprey Publishing Limited
27A Floral Street, London WC2E 9DP
Member company of the George Philip Group
First reprint autumn 1984
Second reprint summer 1985
Third reprint autumn 1985
Fourth reprint spring 1988

British Library Cataloguing in Publication Data

Foale, Tony
  Motorcycle chassis design.
  1. Motorcycles
  I. Title   II. Willoughby, Vic
  623.2'3   TL440
ISBN 0-85045-560-X

Editor Tim Parker
Illustrator Richard Geiger

Filmset and printed in England by
BAS Printers Limited, Over Wallop, Hampshire

Contents

Preface

To the layman, if not the professional, engine design may seem the most glamorous aspect of motorcycle engineering. Yet the design of the chassis is certainly no less important, since it is crucial to handling and safety. To some extent the two aspects are complementary, for the size and shape of the engine may determine the best way for the frame designer to tackle his problems.

In response to general interest, there has never been a dearth of books on engine design and its underlying principles. To the best of our knowledge, however, no comparable work has been published on motorcycle chassis design and construction. With interest increasing, as shown by the continuing demand for alternative frames for both road and racing use, the purpose of this book is to fill that gap, enlighten the layman and stimulate thought on future advances.

We hope the book will also appeal to college students undertaking chassis projects—and will prove a useful guide to their tutors who, while having a sound grounding in engineering, may not be fully conversant with this specialized subject.

To ensure the broadest appeal the book is a compromise between a full technical treatise and a readable volume for the average enthusiast. The trendy jargon beloved by some specialist magazines is deliberately avoided because it inhibits proper understanding and gives rise to many common fallacies.

We have tried to remove the mystique surrounding the subject and show that it is amenable to systematic analysis. Some specialist areas—such as motocross, trials and speedway—are too diverse to be covered specifically but the general principles expounded here may still apply.

Superficially, frame design seems to have progressed little since low-powered engines were first tacked on to bicycles. In some ways, however, development is diversifying— though not always in the right direction and often without a basis in sound principles. The state of the art, so far as the major manufacturers are concerned, is shown by the proliferation of specialist firms in Europe supplying alternative frames for Japanese engines. In any case, the designer is constrained by commercial as well as functional considerations, for the motorcycle market is essentially conservative.

Tony Foale, Tonbridge
Vic Willoughby, Enfield
January 1984

Function and history

Function

The functions of a motorcycle frame are of two types: static and dynamic. In the static sense it is obvious that the frame has to support the weight of the rider or riders, the engine and transmission, and the necessary accessories such as fuel and oil tanks. Although less obvious, the frame's dynamic function is crucially important. In conjunction with the rest of the rolling chassis (i.e. suspension and wheels) it must provide precise steering and good roadholding, handling and comfort.

For precise steering the frame must resist twisting and bending sufficiently to keep the wheels in their proper relationship to one another regardless of the considerable loads imposed by power transmission, bumps, cornering and braking. By proper relationship we mean that the steering axis must remain in the same plane as the rear wheel so as to maintain the designed steering geometry in all conditions without interference from frame distortion.

Clearly, however, no steering system can be effective while the wheels are airborne, hence the importance of roadholding, especially at the front.

Good handling implies that little physical effort should be required for the machine to do our bidding, so avoiding rider fatigue. (This requirement is largely a function of centre-of-gravity, height, overall weight, stiffness and steering geometry.)

Comfort has the same aim (minimum fatigue) and requires the suspension to absorb bumps without jarring the rider or setting up a pitching motion.

All these criteria the frame has to fulfil for the expected life of the machine, without deterioration or failure and without the need for undue maintenance. It should be borne in mind, however, that all design is a compromise. In any particular example, the precise nature of the compromise will be governed by the use for which the machine is intended, the materials available and the price the customer is prepared to pay.

History

Over the years designers have been repeatedly criticized for their seeming reluctance to depart from the diamond-pattern frame inherited from the pedal cycle. However, since the earliest motorcycles were virtually pushbikes with small low-powered engines attached at various places, that was the logical frame type to adopt, particularly so long as pedal assistance was required.

Until the general adoption of rear springing several decades later, this diamond ancestry (including its brazed-lug construction) was discernible in most frame designs. This was hardly surprising as the frame's depth suited the tall single-cylinder engines that were popular for so long; in any case the motorcycle, like the pedal cycle, was after all a single-track vehicle in which the use of an inclined steering head was a convenient way

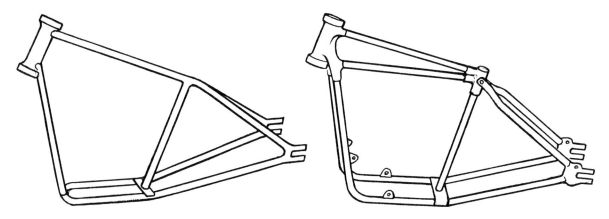

Cradle frame, successor to the diamond pattern. The cradle tubes are extended rearward to the wheel spindle lugs

In the duplex cradle frame the cradle tubes are also extended upward to the steering head lug

to provide the front-wheel trail necessary for automatic straight-line stability.

Once pedals were discarded, the frame with the closest resemblance to the ancestral pedal type was the simple diamond pattern in which the engine's crankcase replaced the cycle's bottom bracket to span the lower ends of the front tube and seat tube. For many years both before and after the first world war this type of frame was the overwhelming choice of the established manufacturers. An earlier variant of the diamond pattern was the single-loop frame in which front tube and seat tube were bent from a single length, which passed underneath the engine.

An improvement on both was the cradle frame. In this, the bottom ends of the single front and seat tubes were spaced farther apart and rigidly connected by a brazed-in engine cradle, from the rear of which the tubes reached upward to the wheel-spindle lugs. A straightforward development of this layout was the duplex cradle frame, in which the cradle tubes were continued upward to the steering-head lug as well as to the rear spindle

lugs. Both types of cradle frame suited the upright single-cylinder engine by providing room for the narrow crankcase to be slung very low; engines with wider crankcases had to be mounted higher, so raising the centre of gravity.

In the design of these early frames, torsional and lateral stiffness seem to have been given a low priority, although a few attempts were made to stiffen the support of the steering head by incorporating it in one end of a cast H-section frame member, which replaced the front down tube in the Greeves and the top tube in some BSAs.

However, there were some commendable efforts between the wars to ensure the all-important torsional and lateral stiffness through triangulation of the frame structure. In the Cotton, the four long tubes connecting the steering head directly to the rear spindle lugs were triangulated in both plan view and elevation, and the machine was renowned for its excellent steering.

Bolted-up from straight tubes (for easy repair) and relying on the power plant for some

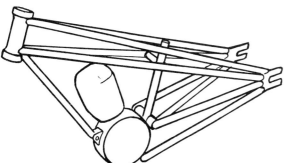

Above The late Brian Stonebridge on a Greeves scrambler with cast light-alloy front down member incorporating the steering head (*MCW*)

Left Triangulated in both plan and elevation, the straight-tube Cotton frame was renowned for its steering

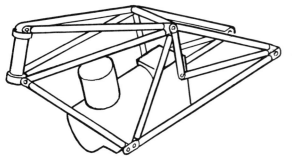

Fully triangulated at the front but only vertically at the rear, this Francis-Barnett frame could be easily repaired by renewing any of the bolted-in tubes

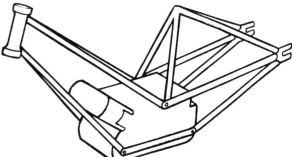

Triangulated fully at the rear but only laterally at the front, this early Scott frame relied on the engine for some of its stiffness

of its stiffness, the Francis-Barnett was fully triangulated from the steering head to the saddle, though the rear end was triangulated only in the vertical plane. Another frame to depend on the engine for part of its stiffness was the open Scott. In this the rear end was triangulated fully but the steering head only laterally—which was much the more important plane from the steering viewpoint. The front brake of the day was hardly powerful enough to tilt the head significantly in the fore-and-aft plane; and even if it did, that would not have impaired the steering nearly so much as would twisting the head sideways. The Scott too earned an enviable reputation for its steering.

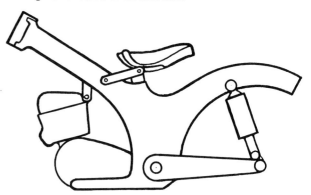

Welded from left and right halves, this NSU beam frame was shaped to accommodate pivoted-fork rear springing and support the engine at the top and rear

Beams

An entirely different approach to the problem of achieving adequate resistance to twisting and bending is to use a large-diameter tube as the main frame member, thus combining a high degree of stiffness with simplicity and light weight. Provided it is of sufficient section, the tube does not necessarily have to be circular, though this is the best shape for torsional stiffness. Indeed, when the NSU Quickly popularized this type of frame at the start of

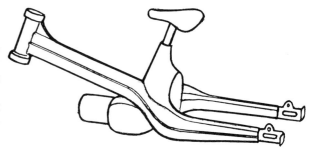

In this unsprung beam frame for a moped the open channel section of the rear fork arms is strengthened by a welded-in U-shape strip

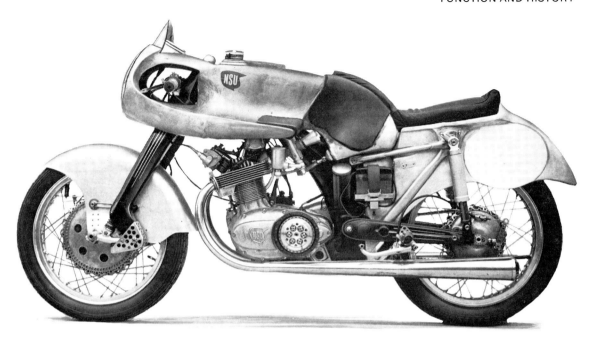

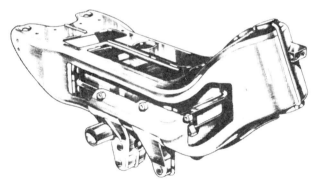

the moped boom in the early 1950s the tube—
or beam, to use another name—was made from
left and right half-pressings seam-welded
together, giving an approximately oval section.

Clearly, however, a plain beam could not
connect the steering head directly to the rear-
wheel spindle as did the top four tubes in the
Cotton frame; hence it was bifurcated at the
rear to accommodate the wheel, and the resulting open channel section of the two arms was
closed by welding-in a U-shape strip to restore
strength.

Welding the beam from two halves in this
way made it possible to incorporate any
necessary curvature in a vertical plane and
NSU used curved beam frames not only on
their moped but also on their Max roadster,
their 250 cc world championship-winning
Sportmax catalogue racing single, and the
works racing 250 cc Rennmax twin and
125 cc Rennfox single.

Above The Ariel Leader (and Arrow) frame derived
great stiffness from the large cross-sectional area
of the beam, which enclosed the 2½-gallon fuel
tank

Top Winner of the world 250 cc championship in
1953, this NSU Rennmax twin had a curved beam
frame welded from left and right halves (*MCW*)

Because of its extremely large cross-sectional area (sufficient to accommodate a separate 2½-gallon fuel tank inside), the Ariel Leader (and Arrow) frame was probably the stiffest and most outstanding of the beam type. Predictably when pressed into racing service, its steering proved well up to the extra demands.

Pivoted-fork rear springing removed the need to bifurcate the rear end of a frame beam because with this type of suspension it is the fork pivot, not the wheel spindle, to which the steering head has to be stiffly connected. (Naturally, the fork itself should continue the torsional and lateral stiffness back to the spindle.)

In the NSU Max and racers, the pressed-steel beam was curved downward at the rear to make a direct connection from steering head to fork pivot. Other frames—such as that on the early grand-prix Hondas and some Reynolds one-offs—used a large-diameter circular tube, similarly curved, to achieve the same effect; the Honda frame, however, like its duplex successor on the grand-prix fours of the early 1960s, was structurally incomplete without the engine, which was attached at the cylinder head and gearbox.

Making a direct connection with a straight tube is usually impracticable, even with a flat-

To make a direct connection between steering head and rear fork pivot, the tubular beam on this early 125 cc Honda grand-prix racer was curved through some 90 degrees (Salmond)

single engine (although the first Foale frame achieved it). Nonetheless, with that type of engine or, say, a sloping parallel-twin two-stroke, the rear end of the tube can be brought within a few inches of the fork pivot, as it was in the Foale frame for a Yamaha TZ350.

In that case, the 2 in. gap was bridged by a small welded-in box section. On a Reynolds frame for a 250 cc Moto Guzzi flat single, however, the gap was appreciably greater and was spanned by a channel-section light-alloy fabrication, bolted through two cross-tubes welded into the main tube (which doubled as an oil tank) and the box-section sump welded to its underside.

A substantially similar layout was used on Norton's Moto Guzzi-inspired experimental 500 cc flat single in the mid-1950s. In that case the oil was contained in a $4\frac{1}{2}$ in.-diameter main tube and a welded-on underslung box that also supported the crankcase, while the

Above The first ever Foale chassis, built in the early 1960s for a small 125 cc engine. Features 3 in. back-bone frame connecting rear fork pivot directly to steering head. Note leading link forks with Greeves type rubber pivot bushes, which also act as the springs

Top A 1975–76 Foale frame for 250 and 350 cc TZ Yamahas. The straight tubular beam is connected to the rear-fork pivot by a box section and stiffened at the steering head by a folded gusset

Somewhat similar to Ken Sprayson's layout, this Norton frame also contained the engine oil in the beam and underslung box, which supported the crankcase. The fork pivot spanned a light-alloy gearbox plate and an aluminium engine casting (*MCW*)

Above Details of the fabrication, which was bolted through cross tubes in the beam and sump and also helped support the engine

Top Reynolds beam frame built by Ken Sprayson for Arthur Wheeler's Moto Guzzi. The beam doubled as an oil tank, with a box-section sump at the rear, and was connected to the rear-fork pivot by a channel-section light-alloy fabrication

fork pivot was bolted between a light-alloy gearbox plate on the left and an aluminium casting on the engine.

When a tall, bulky engine (such as a 1-litre twin-cam four abreast) has to be accommodated, an even bigger gap has to be spanned. A self-defeating scheme adopted by some frame builders was to bridge the gap with a pair of bolted-on light-alloy plates, which could make nonsense of the tube's torsional stiffness, depending on detail design. A much better arrangement was incorporated in the Foale frame for Honda and Kawasaki fours, in which a pair of tubular triangles splayed out from the rear of the tube to the sides of the fork pivot, so providing triangulation in both planes.

Ner-a-Car

Although its chassis comprised two full-length, channel-section sides in pressed steel, cross-braced front and rear, the Ner-a-Car of the 1920s defies classification with the beam-

Left A Foale beam frame designed to accommodate a tall Japanese four-cylinder engine. Actuated by the top apex of the triangulated rear fork, the suspension strut is slung beneath the beam

Below Here the frame is shown with a 1000 cc Kawasaki Z1R engine installed; the tension stays from the steering head support the weight of the engine

In this 1920s Ner-a-Car the chassis comprised left and right channel-section steel pressings, cross-braced front and rear. The engine was slung very low and hub-centre steering was a standard feature

type frames if only because it lacked a conventional steering head to connect to the rear-wheel spindle. Indeed, the steering kingpin was set in the middle of the front axle and housed within the front hub—hence the term: hub-centre steering.

The axle itself was horizontal, shaped like a U (closed end forward) and pivoted in lugs protruding downward from the chassis sides, with stiff coil springs providing a short suspension travel, which thus varied the kingpin inclination. The chassis members were bowed outward at the front for tyre clearance on full lock, which was nonetheless severely restricted.

Although the chassis' resistance to bending in a horizontal plane must have been high, its torsional stiffness was doubtful and the Ner-a-Car's quite exceptional stability most likely stemmed from its ultra-low centre of gravity, allied to an uncommonly long wheelbase— 59 in. for the unsprung version, $68\frac{1}{2}$ in. with quarter-elliptic, pivoted-fork rear springing. Hub-centre steering, however, is more tolerant of torsional flexure.

Stressed engine
One of the earliest examples of the use of the engine as an integral part of the frame was the P & M (later the Panther). In this the very tall

Opposite In old-fashioned riding gear, Vic Willoughby demonstrates the outstanding stability of a Blackburne-engined Ner-a-Car—a characteristic helped by the low centre of gravity and long wheelbase as well as the steering arrangement (*MCW*)

Right **Diagrammatic layout of the stressed-engine principle immortalized by the Series B, C and D Vincent-HRD big twins. The rectangular-section beam connecting the steering head to the cylinder heads contained the engine oil**

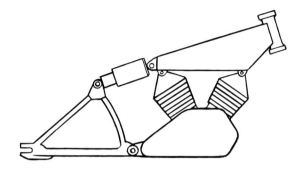

cylinder did duty as a front down tube but the cylinder barrel and head were relieved of tensile stresses by long bolts extending from the cylinder head to the crankcase, or long U-bolts reaching down to loop under the main-bearing housings.

Altogether more renowned and successful was the layout of the postwar Vincent V-twins, in which a rectangular-section backbone, welded from sheet metal in the form of a six-pint oil tank, was bolted to the steering head and both cylinder heads. The triangulated rear fork was pivoted behind the gearbox, while the spring units were anchored to the rear end of the backbone.

Below **A series C Vincent-HRD Rapide with hydraulically damped Girdraulic front fork and twin rear shock absorbers**

Front suspension

The first end of a motorcycle to be sprung was the front, for obvious steering reasons. And, following considerable variety in the early days, the first type of fork to achieve almost universal adoption was the girder, first with side springs then with one spring in front of the steering head. This spring was usually, though not always, of barrel shape to give a progressive rate and easy end fixing.

Amply stiff torsionally, many girder forks nonetheless lacked intrinsic lateral stiffness, though some notable attempts were made to

Left to right **Girder, telescopic, leading-link and trailing-link front forks, showing the path travelled by the wheel spindle throughout suspension movement**

improve this. Both Rudge and Vincent (on the Girdraulic fork) stiffened the link assemblies by forging the links integrally with the spindle housing (the Girdraulic also had forged light-alloy fork blades). And the Webb fork fitted to some early KTT Velocettes had the girders triangulated laterally by tubes joining the middle of the bottom spindle housing to lugs outboard of the fork ends.

After a long reign, the girder fork eventually gave way to the hydraulically damped telescopic type so popular today when BMW demonstrated the latter's superior characteristics in classic racing from 1935 onward. Compared with the girder fork, the telescopic required no adjustment or routine lubrication, provided longer wheel travel, gave substantially constant trail under most conditions (except for nose-diving under braking, when the trail reduces) and had much superior damping characteristics.

With the exception of snubber springs, pioneered on the works Norton racers of the late 1930s, most girder forks relied on friction damping. Unfortunately, the characteristics were opposite to those required—i.e. resistance to initial movement (stiction) was too stiff and it softened considerably once movement started. By contrast, hydraulic damping is automatically related to the *rate* of wheel deflection; and, unlike friction damping,

it does not have to give the same resistance in both directions; hence it can easily be tailored to most requirements.

Yet BMW were by no means first with a telescopic front fork. Several decades earlier Scott employed the principle, though without damping. Nevertheless, their fork was well engineered, with the sliders firmly braced by being brazed into a lug above the wheel (where they operated a central spring) and with the bottom bushes well supported, latterly by diamond-shaped girders.

Although the hydraulically damped telescopic fork quickly matched the earlier girder in being almost universally adopted, its instant popularity probably owed more to its neat appearance, low production cost and lack of maintenance than to its dynamic qualities. Despite its superiority to the girder pattern many examples lacked sufficient torsional and lateral stiffness, especially for racing. And its hydraulic damping, although a great advance on the friction type, was not so refined as that in the proprietary struts supplied by specialist concerns for rear springing. (In the mono-coque Norton on which he won the 1973 Formula 750 TT, Peter Williams showed what could be achieved here by incorporating Koni damper units in the legs of a telescopic fork; for series production, however, the extra cost might well have been considered unacceptable.)

Where cost had a lower priority—i.e. in grand-prix racing—or where a manufacturer of roadsters set more store by high quality than low price and sleek looks, some engineers rejected the telescopic fork for its dynamic and structural shortcomings and preferred the leading-link variety. This offered greater tor-sional and lateral stiffness, lower unsprung weight, better damping (through proprietary struts) and a choice of steering geometry between virtually constant trail and constant wheelbase, depending on the inclination of the links.

For world-championship racing at the highest level, both NSU and DKW discarded telescopic forks in favour of leading links while

Manfred Schweiger's 1929 Scott TT Replica, showing one of the earliest telescopic front forks, in which the bottom bushes were supported by diamond-shaped girders. The lug at the top of the sliders actuated a central spring (Magrholz)

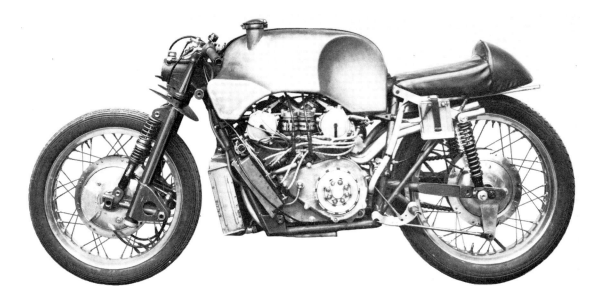

BMW changed to an Earles-type pivoted fork. In the same context, Moto Guzzi never seriously considered anything but leading links; neither did Reynolds. For both road and racing use, Foale made a similar choice; so did Greeves, Cotton and others.

Two methods have been used to combine lightness with rigidity. Some designs (e.g. Greeves and Foale) united the left and right links with a tubular loop round the back of the wheel; Moto Guzzi, however, whose racing machines were unsurpassed for handling,

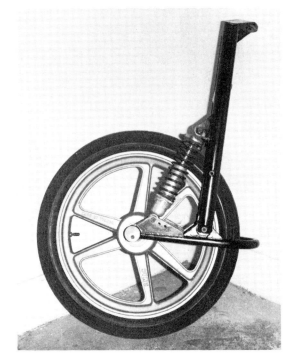

Top One of the most famous leading-link front forks—that on the 1956 version of the 500 cc Moto Guzzi V-eight grand-prix racer. The suspension struts were externally mounted for easy experimentation (*MCW*)

Right A Foale leading-link front fork with the links formed by a loop behind the wheel. Alternative bottom locations for the suspension strut permit variations of the effective spring and damping rates; eccentric spindle mounting allows trail changes

achieved ample stiffness without a loop by using a large-diameter (hollow) wheel spindle secured in the links by wide clamps, each with four nuts.

Given the same inclination of the links, a trailing-link fork may have similar characteristics to those of a leading-link, except for higher steering inertia. Yet the difficulty of making a sound and neat job of its design discouraged most manufacturers from trying it. Ariel were the most notable exception, using trailing links on the Leader and Arrow, both of which were noted for impeccable handling.

A variant of the leading-link fork was the so-called Earles type, with long links combined into an integral fork pivoted behind the wheel. BMW standardized this arrangement for practically two decades between abandoning their original telescopic fork and introducing the long-travel version in 1969. Essentially, though, their Earles-type fork was a compromise for solo and sidecar duty. And although its lack of sliding friction was a boon in sidecar cornering, its high steering inertia blunted steering response in a solo.

Rear suspension

In general adoption, rear springing lagged several decades behind the sprung front fork, largely because of the long dominance of the rigid frame in road racing and the inevitable poor handling of some early bolt-on conversion kits.

Plunger springing was first to gain wide acceptance—partly because of racing successes by BMW and Norton, and partly because,

This 1956 Senior TT shot of Walter Zeller shows the Earles-type pivoted front fork on the racing BMW. The brake shoe-plate was clamped directly to the left fork arm to resist diving. The telescopic steering damper can be clearly seen (*MCW*)

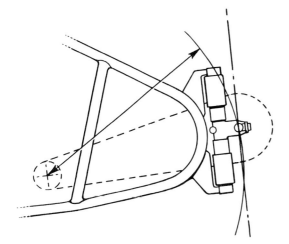

Left **With plunger-type rear suspension the straight-line movement of the wheel tightens the chain at the extremes of travel, thus setting a limit to total movement and giving a slack chain at static load**

for the manufacturer, it was the easiest type to which existing rigid frames could be adapted. However, its limitations were clear from the start. First, the incorporation of spring boxes at the rear destroyed the triangulation of the seatstays and chainstays, thus permitting each side to flex independently in a vertical plane, even to the extent of fatigue failure. Second, resistance to wheel tilting depended also on very stiff clamping of the spindle in the sliders

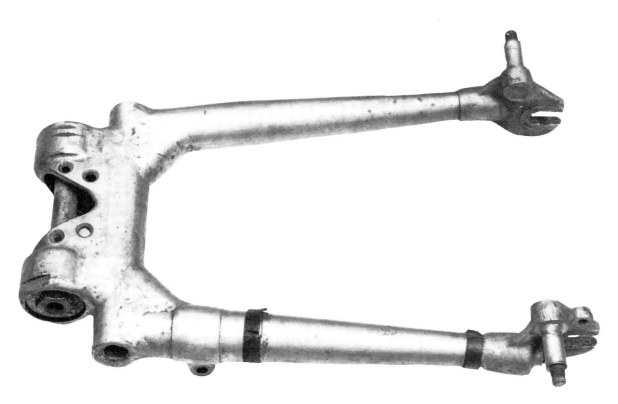

(as it does with telescopic front forks and some leading links). Finally, the straight-line movement of the wheel tightened the chain considerably at the extremes of travel, so limiting the range of wheel movement and necessitating a slack chain setting in the static load position.

With such notable exceptions as AJS/Matchless, most manufacturers found their existing rigid frames unsuitable for adaptation to pivoted-fork rear springing. Even so, that form of springing was soon recognized as being superior to the plunger variety on all other counts. Indeed, it preceded plungers, with NSU and Indian providing early examples about the time of the first world war, and with Vincent-HRD standardizing their famous pivoted-fork layout from the outset in 1928 until production ceased in 1955.

Vincent obtained great strength and rigidity by triangulating the arms of the fork, mounting it on a very wide pivot and using preloaded taper-roller bearings to obviate play. Moto Guzzi also chose to triangulate the fork when they introduced rear springing in 1935 and scored spectacular victories in the Senior and Lightweight TTs through Stanley Woods. Later on, however, they changed to a plain fork welded from very large diameter tubing and claimed it to be equally stiff torsionally and stiffer laterally (because the earlier fork was triangulated only vertically, not laterally).

Another pivoted rear fork designed to provide ample stiffness without triangulation was introduced by Velocette on the works racers of the mid-1930s and standardized on the Mark VIII KTT in 1939. In this, one of the heaviest layouts, the arms comprised taper-diameter,

Left **In a bid to provide ample stiffness without triangulation the Velocette pivoted rear fork was made from taper-section, taper-gauge tubing and cross-braced behind the pivot. Here the tubes are adapted to a Vincent lug for George Brown's world record twin**

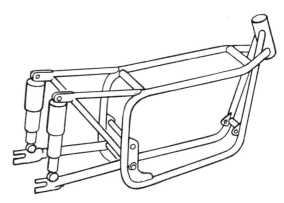

The so-called featherbed Norton duplex-loop frame with plain pivoted rear fork. The crossing over of the tubes at the steering head facilitated the use of a flat-bottom tank but impaired stiffness; a head steady was added which connected to the steering head to overcome this

taper-gauge tubes, cross-braced just behind the widely spaced pivot bushes, which were supported on a large-diameter hollow spindle pressed into a robust lug on the vertical seat tube.

But many plain forks lacked adquate torsional stiffness and this gave rise to a vogue for precisely matched left and right suspension struts to minimise one of the causes of twisting.

Probably the most renowned frame with a plain pivoted fork was the so-called Norton featherbed introduced on the works racers in 1950. Its enormous improvement over its predecessor (the plunger-sprung 'garden gate') probably owed less to any one design feature than to a combination of several. Its duplex-loop structure provided adequate (though not exceptional) stiffness and the general layout was such as to give a fairly even weight distribution and a relatively low centre of gravity (considering the upright position of the cylinder); the front fork was one of the more robust telescopics; and the steering geometry provided light, responsive handling.

This view of a so-called garden-gate Manx Norton shows how plunger-type rear springing destroyed the triangulation of the chainstays and seatstays in the unsprung frame it replaced. At the front, wheel location is by means of the highly regarded 'Roadholder' hydraulically damped telescopic forks

Opposite, top On Doug Hele's **1952 BSA 250** cc grand-prix racer the top apex of the triangulated rear fork actuated a horizontal suspension strut in a tunnel through the oil tank. Front fork was leading link. Note steered headstock arrangement (*MCW*)

Opposite, bottom Rear springing on Kenny Roberts' **1978** works **500** cc Yamaha four. Though given the fancy names of cantilever and monoshock, it was in principle no different from the Vincent Series D layout. Triangulated pivoted fork was in light alloy (*MCW*)

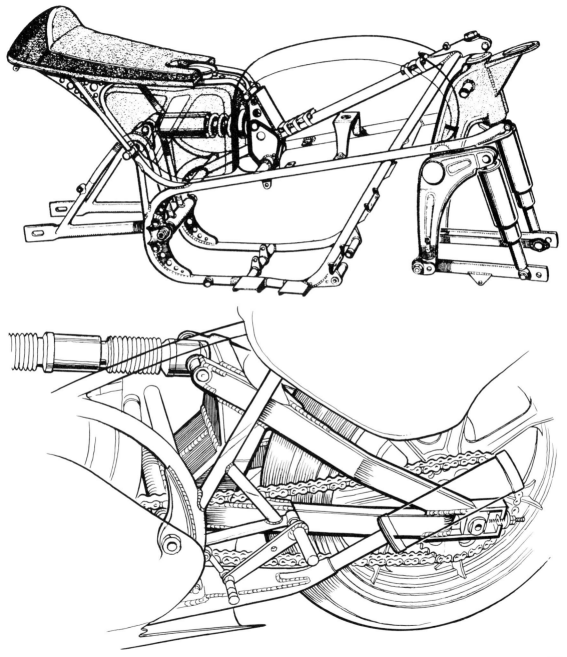

There have been various permutations on rear fork types and spring struts. A fork triangulated above pivot level lends itself to a single strut actuated by the top apex of the fork, as on the experimental 250 cc BSA grand prix racer built by Doug Hele in 1952. For their Series D twins in 1955 Vincent switched to a single Armstrong strut in place of the previous two spring boxes and separate damper under the seat. Nearly two decades later Yamaha revived the system, which then acquired the fancy names of cantilever and monoshock.

Moto Guzzi's original sprung fork was triangulated below pivot level to keep the weight low down. At first it was linked to a pair of long horizontal spring boxes flanking the

rear wheel; later on these gave way to a single strut beneath the engine, anchored at the rear and compressed from the front by a rod passing through the spring. The latter arrangement was retained when the triangulation was abandoned.

For several years on the RG500 racer, Suzuki also triangulated the rear fork below pivot level, yet retained a pair of struts above the fork. Then, on their 250 cc grand-prix twins in

A fairly modern example of a rear fork triangulated below pivot level, on the Suzuki RG500 racer. Unlike the original Moto Guzzi layout of this type, two suspension struts are fitted, in the conventional position

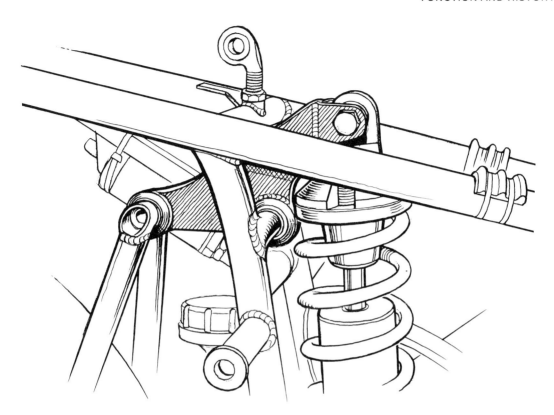

the mid-1970s, Kawasaki started a racing trend towards rocker-arm rear suspension (though the system was already used in motocross). In the original layout the single suspension strut was installed vertically behind the gearbox and anchored to the frame at the bottom end. Fork movement was transmitted to the top of the strut through a short rocker arm linked by an A-bracket to points midway along the fork arms.

In a development of the system for the Kawasaki KR500 square four, the fork was triangulated above pivot level and its apex connected to the rocker by a short upright link; the strut was no longer anchored to the main frame but to the fork itself, just behind and the below the pivot.

One of the earliest rocker-arm rear-suspension layouts, on Kork Ballington's 250 cc Kawasaki grand-prix machine. The A-bracket on the left was attached to the pivoted fork while the suspension strut was anchored to the frame at the bottom (*MCW*)

In both cases the chief aim was to achieve a progressively stiffer resistance to wheel deflection (while using a constant-rate spring) so combining sensitivity to small bumps with an increasing check for larger ones. Although the resistance offered by the spring itself was directly proportional to its own deflection, the *effective* rate, measured at the wheel spindle, depended on the angles in the linkage.

Though the principle was widely adopted, the geometry was sometimes suspect, with the

29

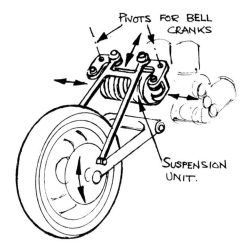

effects of some changes in linkage angles tending to cancel out. Some designers eschewed complex linkages and relied instead on struts with variable-rate springs—e.g. either two different springs end to end or one spring wound to different pitches.

Perhaps the most ingenious of the rocker systems was that on Yamaha's OW61 grand prix four. In this, the suspension strut was mounted transversely and squeezed from both ends simultaneously by bell-cranks linked to the top corner of the triangulated fork. The main advantage of this layout seems to be fore-and-aft space saving.

Above Space-saving rear-springing layout on Yamaha's OW61 grand-prix four. The suspension strut was squeezed, concertina-fashion, by two bell-cranks linked to the top apex of the triangulated fork (*MCN*)

Opposite Boge Nivomat installation on an 850 cc Moto Guzzi California. Regardless of load, this automatically keeps ride height and suspension travel constant

Below Prior to the advent of Boge Nivomat self-levelling suspension struts, the best means of compensating for different loads was this Velocette arrangement of slotted top anchorages, which altered both spring and damping effective rates

Other spring types

From time to time rubber and air have been used as the springing medium instead of steel. Rubber in tension, in the form of bands, has been used in both telescopic and leading-link forks and by Tony Foale for rear springing. On some Greeves machines, rubber was used in shear in the form of bonded bushes at the pivots of the fork links. Although rubber can be endowed with varying degrees of inherent damping, this was never sufficient to preclude the use of separate friction or hydraulic dampers.

A significant advantage of air springing (as also of rubber) is its progressive rate, which can be demonstrated by operating a tyre pump with the outlet sealed by one finger. A possible disadvantage, for really extreme conditions, is that the air heats up, so increasing its pressure, hence its effective rate.

Velocette introduced this form of springing, made by Dowty, on their works racing machines in the mid-1930s, subsequently extending its use to the rear end of the KTT in 1939 and the front end of some of their post-war roadsters.

In production, leakage of both air and damping oil was a problem. But improvements in sealing materials and surface finishes have enabled the Japanese to revive air springing, first as an auxiliary to steel springs, then on its own. On the whole, however, single- or multi-rate coil springs remain the first choice for most designers.

Compensation

Providing ready compensation for different weights of rider and the added weight of a pillion passenger has always bedevilled the design of rear springing for roadsters because the load on the springs may easily be increased by 50 or even 100 per cent. Under conditions of varying load, multi-rate springing, however achieved, may be an advantage.

The most common solution to the problem has been to provide a stepped cam adjustment for the preload on the springs. By this means the attitude of the machine can be preserved for different loads and full travel retained. But no single spring rate can be satisfactory; in any case, only BMW and MZ provided a hand adjustment for the pre-load and few riders bothered to use the C-spanner otherwise necessary.

A much better form of compensation was that designed by Phil Irving for Velocette. In this, the top anchorage for the struts was slotted so that their inclination could be varied to suit the load by slackening two handwheels. Thus both the effective spring rate and the damping, measured at the wheel spindle, were adjustable; but in later production versions a spanner was required and so the adjustment was used less than it deserved to be. A consequence of this scheme is that less wheel movement is available on the hard setting than the soft.

A later solution to this problem was the Boge Nivomat self-levelling strut first offered on BMW roadsters in the early 1970s.

Basic principles

Balance

As a single-track vehicle, a motorcycle lacks inherent static balance; hence it needs supporting if it is to remain upright when stationary. Once it is moving, however, it should balance itself from as low a speed as possible.

At really low speeds, when the gyroscopic effects of the wheels are negligible, balance depends mainly on the rider's skill, the geometry of the machine, the position of its centre of gravity and its weight. At higher speeds, balance is less reliant on the rider's skill and is due mostly to the gyroscopic forces of the wheels.

A gyroscope, which is basically a wheel spinning at high speed, has a very stable axis— i.e. a strong tendency to maintain its angular plane of rotation. In other words, while it can easily be moved laterally (i.e. along its axis), it resists tilting in any direction; and when it *is* tilted it automatically twists strongly in a plane at 90 degrees to the plane of tilt. These forces increase rapidly as the speed of rotation rises.

The wheels of a motorcycle act as gyroscopes (as also does the engine crankshaft), thus tending to keep the bike upright and in a straight line—in a similar manner to that of a spinning top, which can balance on a point when spinning fast but topples over when rotating slowly or, of course, when stationary.

The response of a gyroscope to tilting is called gyroscopic precession; and a few experiments with a bicycle wheel will not only prove the strength of the effect but will also show the subtleties that contribute to balance and steering.

First you should hold the wheel upright by the spindle ends and get someone to spin it briskly so that the top of the wheel is moving away from you, as if it were the front wheel of a machine you were riding. If you then try to tilt the spindle to the left (equivalent to banking your machine) you will find the wheel turns instantly and strongly to the left, as if steered by an invisible hand. In other words, your attempt to tilt the wheel about its fore-

Fig 2.1 **Gyroscopic precession. When a bicycle wheel spinning as shown is turned to the left, it tilts strongly to the right; when it is tilted to the left it turns to the left**

33

and-aft horizontal axis has produced a couple swivelling it about its vertical axis.

Now start again but this time try to steer the spinning wheel to the left while keeping it upright. Just as sharply and strongly it will bank to the *right*.

We shall see later that the direction in which the couple operates is important for balance and steering. And it is equally important to note that the strength of this gyroscopic precession depends not on the extent to which you tilt or turn the spindle but on how sharply you do so.

Let us now see how these forces automatically keep the machine balanced and on a straight path without assistance from the rider. Suppose the machine, while travelling at a normal speed, starts to fall to the left under some extraneous influence. As we have just seen, gyroscopic precession of the front wheel immediately turns it to the left. This sets the machine on a curved path, so creating a centrifugal force which counteracts the lean and restores the machine to the vertical; the effect of the gyroscopic precession is reversed and returns the steering to the straight-ahead position.

Steering

So much for balance. Our next problem is steering and, for the purpose of analysis, we can divide this into two phases: 1) initiating the turn; 2) maintaining the turn.

Since the second phase is easier to analyse, let us consider it first. It is not feasible to steer a solo motorcycle through a corner in a substantially upright position (as it is with a sidecar outfit or car) because the centrifugal force generated would cause it to fall outward. Hence we must bank the machine inward so that this tendency is counteracted by the machine's weight tending to make it fall inward.

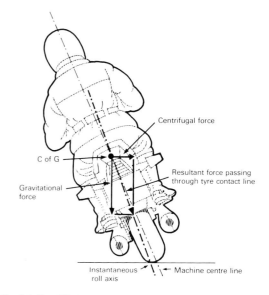

Fig 2.2 **Equilibrium in cornering is achieved when the resultant of centrifugal force and gravitational force (both acting through the mass centre) passes through the line joining the contact patches of the front and rear tyres**

Equilibrium is achieved when the angle of lean is such as to balance the two moments—one due to centrifugal force acting outward and the other to gravitational force acting downward (both through the mass centre). The actual angle, which depends on the radius of the turn and the speed of the machine, is that at which the resultant of the two forces passes through a line joining the front and rear tyre contact patches. (This is the roll axis.)

But how do we initiate the turn—do we lean first or steer first? If we were to turn the handlebar in the direction we want to go, both centrifugal force and gyroscopic precession of the front wheel would cause the bike to topple outward. But if we momentarily try to turn the bar in the opposite direction, then the centrifugal and gyroscopic forces will cause the machine to bank to the correct side. Gravity will then augment the banking effect and this, in turn, will give rise to gyroscopic forces helping to steer the front wheel into the curve.

These forces will also act on the rear wheel which, because it is rigidly attached to the bulk of the machine, will tend to make the machine yaw into the curve. However, this reinforcing effect is secondary to that of the front wheel. (Steering rake and front-wheel trail, i.e. castor, also help steer the machine into the curve, as we shall show later.)

We have seen, then, that a turn can be initiated by steering momentarily in the 'wrong' direction. But that doesn't explain how we can corner no hands. So let us consider what happens if we try to lean first.

As we saw earlier, simply banking the bike steers the front wheel in the right direction automatically through gyroscopic precession (and the trail effect just mentioned). But as there is nothing solid for us to push against, the only way we can apply bank is to push against the machine with the inertia of our own body. To lean the bike to the left, we must therefore initially move our body to the right.

We now have two possible methods of initiating a turn and it is interesting to note that in both of them (banking and reverse handlebar torque) our physical effort is in the opposite sense to what might be thought natural; but when learning we adapt quickly and the required action becomes automatic.

In practice we subconsciously combine both methods and the pressure on the inner handgrip is partly forward, partly downward. Remember, though, that the forward movement (reverse steering torque) is very small, since gyroscopic precession depends for its strength on the sharpness of the movement, not its extent.

The relative proportions in which we combine the two methods depend partly on riding style but also on speed and machine character-istics. For example, a heavy machine with light wheels at low speeds demands a different technique from that appropriate to a light machine with heavy wheels at high speeds and hence a different feel. But adaptation is quick and soon becomes second nature.

The reason we have not so far dealt with such important matters as steering geometry, wheel and tyre size, wheelbase, centre-of-gravity height, frame stiffness and so on, is simply that balance and the ability to initiate and maintain a turn can be achieved within a wide range of these parameters, which we shall now examine in detail. First, some definitions.

Handling

By this we mean the ease, style and feel with which the motorcycle does our bidding. It depends mainly on steering geometry, chassis stiffness, weight and its distribution, tyre type and size.

Roadholding

This means the ability of the machine, through its tyres, to maintain contact with the road. It depends mainly on tyre type and size, suspension characteristics, weight and its distribution, and stiffness between the wheels to maintain their correct relationship to one another.

In the days of relatively narrow tyres, roadholding and handling generally went hand-in-hand; indeed, the terms were used interchangeably. But nowadays the requirements are sometimes contradictory and a compromise must be struck, depending on the intended use of the machine.

Stability

There are formal definitions for this but they are too involved for a book of this nature. For our purposes we mean: a) the ability to maintain the intended manoeuvre (i.e. continue in a straight line or round a corner) without an inherent tendency to deviate from our chosen path; b) the ability to revert to the intended manoeuvre when temporarily disturbed by external forces (e.g. bumps, cross winds and so on).

Handling, roadholding and stability are affected by many parameters and the interaction between them. The subject is complex and—judging from some chassis designs—not always well understood. Let us, then, consider the main parameters involved and study their various effects. It must be emphasized that there is much cross-coupling between the effects—there is no 'correct' combination, no 'perfect' design. Any motorcycle embodies several essential compromises.

Steering geometry

The basic elements of this are shown in figure 2.3.

Trail

The primary function of this is to build in a certain amount of straight-line stability. We can see that both front and rear tyres contact the ground behind the point where the steering axis meets it; this gives rise to a castor (self-centring) effect on both wheels. The measurement of this castor (steering axis to centre of contact patch) is called the trail.

How the trail causes a self-centring effect can be understood from figure 2.4, which is a plan view of a wheel displaced from the straight-ahead position. Because the wheel is at an angle to the direction of travel (slip angle is the technical term) a force at right-angles to the tyre is generated. Since the contact patch is behind the steering axis (positive trail) then this force acts on a lever arm (approximately equal to the trail) to provide a correcting torque to the angled wheel. That is to say, if the steering is deflected by some cause (e.g.

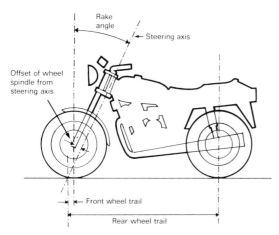

Fig 2.3 **Rake is the rearward inclination of the steering axis; trail is the amount by which the centre of the tyre contact patch trails behind the point where the steering axis meets the ground. Offset (of wheel spindle from steering axis) is measured at right angles to the axis**

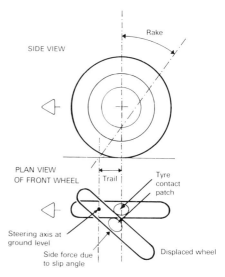

Fig 2.4 **How positive trail and the side force due to slip angle combine to restore deflected steering to the straight-ahead position**

uneven road surface) then positive trail auto-matically counteracts the deflection and so gives a measure of directional stability. This supplements the stabilizing effects of the gyro-scopic forces.

If the tyre contact patch was in front of the steering axis (negative trail) then the torque generated would reinforce the original disturb-ance and so make the machine directionally unstable.

One may be forgiven for thinking that, because the positive trail of the rear wheel is much greater than that of the front (typically 2–4 in. front, 55–60 in. rear) the rear wheel is the more important in this respect. The reverse is the case and there are several reasons for this. See figure 2.5.

Imagine that the contact patch of each wheel is in turn displaced sideways by the same amount (say $\frac{1}{2}$ in.). The front wheel will be turned approximately 7–10 degrees about the steering axis; this gives rise to a slip angle of the same amount and generates a sideways force that has only the relatively small inertia of the front wheel and fork to accelerate back to the straight-ahead position.

But the slip angle of the displaced rear wheel will be much less (approximately $\frac{1}{2}$ degree) and so will the restoring force. Not only do we have a smaller force but it also has to act on

the inertia of a major proportion of the machine and rider, hence the response is much slower than in the case of the front wheel.

From this, we can see that increasing the trail as a means of increasing the restoring tendency on the wheels is subject to the law of diminishing returns. It must also be emphasized that the disturbance to a machine's direction of travel, due to sideways displacement of the tyre contact patch, is less from the rear wheel than the front because of the much smaller angle to the direction of travel that the displacement causes.

To summarize we can say that, while the large trail of the rear wheel has a relatively small restoring effect, the effect of rear-wheel dis-placement on directional stability is also small and so compensates.

Although the primary purpose of front-wheel trail is to provide a degree of directional stability, there are various side effects too; let us consider two of them.

a) Steering effect
If we lean a stationary machine to one side and then turn the handlebar the steering head rises or falls, depending on the position of the steer-ing. In motion, the effective weight of bike and rider supported by the steering head (this weight is increased when cornering by centrifugally-induced forces) is reacted to the ground through the tyre contact patch. This force tends to turn the steering to the position where the steering head is lowest (i.e. the posi-tion of minimum potential energy). For a given

Fig. 2.5 **Despite the much larger trail of the rear wheel, its slip angle for a given lateral displace-ment is considerably smaller than that of the front wheel; hence both the upsetting effect of the displacement and the restoring force are less significant**

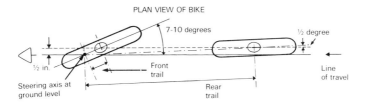

amount of trail, this steering angle is affected by rake and wheel diameter, as discussed later.

If we have positive trail, as is normal, then the turning effect is into the corner (it would be the other way for negative trail). Thus the amount of front-wheel trail affects the amount of steering torque the rider must apply (hence the feel of the steering) to maintain the correct steering angle consistent with the radius of the turn and the machine's speed.

b) Straight-line feel
As we all know, even when we are riding straight ahead the steering feels lighter on wet or slippery roads than on dry. This is because our seemingly straight line is actually a series of balance-correcting curves, with the handlebar turning minutely from side to side all the time. As we have seen earlier, a small steering displacement causes a tyre slip angle, which produces a restoring torque. For a given slip angle, this torque depends on tyre properties, surface adhesion and trail. Thus, through the steering, we get a feedback on road conditions that gives us a feel (dependent on trail) for the amount of grip available.

Rake angle (steering axis angle)
The basic reason for rake is less easy to explain than that for trail because it is just possible that no rake angle at all would be best. Why, then, do all current production machines have the steering axis raked between 25 and 30 degrees from the vertical—usually the 'magic' 27 or 28 degrees? There is no simple answer and several factors are probably relevant:
a) Convenience of construction (see figure 2.6).
b) Lack of imagination.
c) Manufacturers' fear that too great a departure from tradition may inhibit sales. The motorcycle market is conservative and several adventurous makers have gone out of business through producing unusual designs.

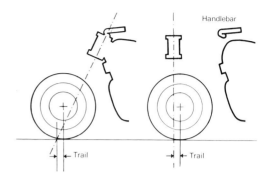

Fig. 2.6 **With a conventional high steering head, a normal rake angle (say, 28 degrees) is convenient for frame construction and direct handlebar mounting. For the same trail, a vertical steering head brings problems in both respects. See page 60**

In most designs of hub-centre steering, simply for space reasons, the wheel spindle is not offset from the steering axis; hence trail is wholly dependent on rake angle, which is typically between 3 and 15 degrees to give the required result. This is a much steeper angle than usual, yet hub-centre layouts are renowned for their stability and steering. Their reputation may stem from reasons other than the steep rake angle; but it certainly seems that this departure from the norm is not harmful and may indeed be beneficial.

Let us now examine the main effects of rake angle. Figure 2.7 shows three different rake angles, all giving the same amount of trail.

1) Reduction of castor effect
For a given amount of trail, the self-aligning torque on the front wheel and fork depends on the length of the lever arm from the centre of the tyre contact patch to the steering axis, measured at right angles to that axis.

As can clearly be seen in figure 2.8, this lever arm is shortened as the rake is increased. In practice, this means that we need more trail as the rake angle is increased (steering head inclined further backward).

Fig. 2.7 **Three rake angles giving the same trail.** *Left:* **Conventional arrangement.** *Middle:* **Rake angle for zero spindle offset (as in many types of hub-centre steering).** *Right:* **zero rake angle (vertical steering axis); note negative offset**

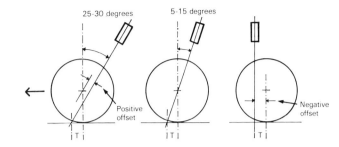

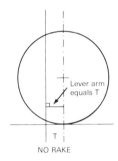

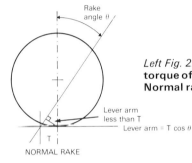

Left Fig. 2.8 **Positive rake reduces the self-centring torque of a given trail dimension.** *Left* **No rake.** *Right* **Normal rake**

2) Negative castor

At large steering angles, the rake can even cause the castor effect to become negative, though sufficiently large angles are possible only at very low speeds.

Reference to figure 2.9 clearly shows what happens. Even though very large steering angles are needed to produce negative castor, there is still some reduction in trail at smaller steering angles and this may necessitate using a larger initial trail than would otherwise be so. (A convincing demonstration of this effect can be obtained by wheeling a pushbike and turning the handlebar far enough for negative castor to take over, whereupon the steering will try to turn even further.)

This is one reason why trials bikes have a steep steering axis, since their tricky manoeuvres at low speeds often involve extreme steerings angles.

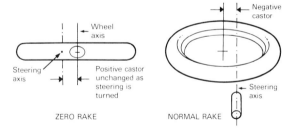

Fig. 2.9 **Plan view with steering turned 90 degrees to the left, showing negative castor with normal rake and retention of full positive castor with zero rake**

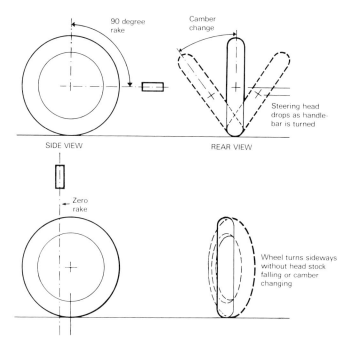

Fig. 2.10 **With a 90-degree rake (top), steering-head drop and camber change on turning the handlebar can be seen. With zero rake (bottom) these effects are absent**

3) Steering-head drop

With a normal motorcycle (i.e. with positive trail) held vertically, the steering head will drop as we turn the handlebar to either side. (With negative trail, which is not normal, it would rise.) The greater the rake angle, the greater the drop. This can be best appreciated by visualizing an extreme rake angle, as in figure 2.10.

This drop tends to work against the self-centring effect of castor because, to return deflected steering to the straight-ahead position, we must lift the considerable weight supported by the steering head. While this effect is detrimental to balance (hence another reason why trials bikes have steep head angles), and to directional stability while

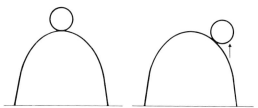

Fig. 2.11 **Analagous to steering head drop is the case of a ball on a mound. Virtually no force is required to keep it in place at the top; if it is allowed to drop, however, we must lift its weight to restore balance. This condition is known as unstable equilibrium**

travelling in a straight line, it helps to steer the wheel into a corner when banked over, thus reinforcing the self-steering effect of trail mentioned earlier.

Centre of gravity of steered parts ahead of steering axis

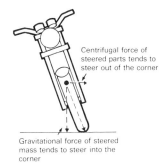

Centrifugal force of steered parts tends to steer out of the corner

Gravitational force of steered mass tends to steer into the corner

Fig. 2.12 The forward offset of the centre of gravity of the steered parts causes turning moments due to gravitational and centrifugal forces which balance one another provided that the rider's centre of gravity is in the machine's centre plane. The racing fashion of leaning in farther than the machine reduces the gravitational moment and increases the centrifugal moment; hence the resultant moment tends to cause a steering effect out of the bend

4) Camber change
Figure 2.10 shows what happens with a 90-degree rake angle—camber change. Although the effect is less pronounced with conventional rake angles, it is still there— and it means that, when a motorcycle is cornering, the front wheel leans over more than the rear.

5) Wheel-spindle offset
Reference to figure 2.7 shows the difference in offset required with different rake angles to achieve a given amount of trail; zero rake angle requires the greatest offset. However, since a normal rake angle tends to reduce the effectiveness of the trail dimension, a zero rake angle would require less trail, hence less offset.

In general, all other factors being equal, it is an advantage to have the minimum offset, since this gives minimum steering inertia. In this respect, hub-centre steering seems to have much in its favour.

It is a fallacy to imagine that, because the offset places the centre of gravity of the wheel and fork ahead of the steering axis, this prod-uces a torque tending to steer the wheel into the curve while the machine is banked over. This is true only when the bike is stationary. During cornering, centrifugal force acts through the offset to steer the wheel *out* of the turn; but this effect is exactly balanced by that of the gravitational force tending to steer into the turn. Hence the offset has no net effect on the machine's self-steering characteristics.

6) Gyroscopic effects
When explaining the automatic balancing effects of gyroscopic precession, we treated the subject as if the steering axis was vertical (i.e. zero rake). In the case of a normal rake angle the situation is modified: the components of the precessional forces acting as described are reduced and components are introduced that act in a contrary way, so reducing the effectiveness of the gyroscopic forces.

7) Steering angle
A rake angle reduces the effective steering angle between tyre and ground compared with the angle through which the handlebar is turned. This can easily be seen by our trick of visualizing the extreme rake of 90 degrees. See figure 2.13. In this case no true steering angle is developed but a camber angle is produced equal to the handlebar angle. At more normal rake angles a small camber change is produced (as in 4 above); for a 27-degree rake, this redu-

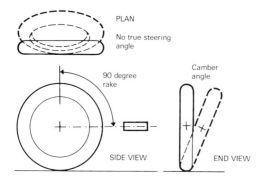

PLAN

No true steering angle

90 degree rake

Camber angle

SIDE VIEW

END VIEW

Fig. 2.13 **This extreme example shows how steering rake reduces the effective steering angle**

ces the effective steering angle to approximately 90 per cent of the handlebar angle.

Summary of rake effects

Except for the case of minimum offset in 5 (above) and the weight-assisted self-steering

effect mentioned in 3, it seems that a non-zero rake angle is no good thing. If this is so, why do conventional motorcycles handle and steer as well as they do? The answer lies in the very small steering angles involved in normal riding. The detrimental effects of rake become more pronounced at greater steering angles.

The weight-assisted self-steering effect may or may not be beneficial—it is possible to have too much of a good thing. When cornering at any particular bank angle and speed we need a self-steering effect to give us just the right steering angle; too much and the rider must apply a reverse effort to the handlebar, too little

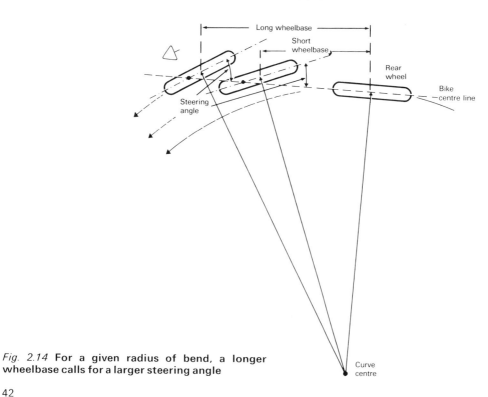

Long wheelbase

Short wheelbase

Rear wheel

Bike centre line

Steering angle

Curve centre

Fig. 2.14 **For a given radius of bend, a longer wheelbase calls for a larger steering angle**

and he needs to steer into the corner.

But since the steering angle required for a given angle of bank varies according to the speed, it is not possible to build in a self-steering effect that is perfect for all speeds and bend radii—which is just another example of the unavoidable need for compromise.

Wheelbase
The distance between the wheel centres has several effects but, in general, the longer the wheelbase the greater the directional stability and the greater the effort needed to negotiate bends. There are three reasons for this.

1) Required steering angle
Figure 2.14 shows how, for a given bend, a long-wheelbase machine needs the front wheel to be turned farther into the bend. Consequently more effort is required for cornering; also a given deflection of the front wheel (say, from bumps) will have less effect on its directional stability.

A practical consideration for trials bikes is that, for a given degree of steering lock, the minimum turning circle is smaller with a shorter wheelbase. For this reason, trials machines have wheelbases as short as 49–50 in.

2) Rear-wheel angle
It is also clear from figure 2.15 that, for a given sideways deflection, the angle of the rear wheel to the direction of travel is smaller with a longer wheelbase, thus improving directional stability.

3) Inertia effects
The wheelbase has an effect on weight transfer under braking and acceleration: for a given centre-of-gravity height, the longer the wheelbase the smaller the weight transfer. Also the moments of inertia in the pitch and yaw planes are increased, which makes a long-wheelbase machine more sluggish and stable.

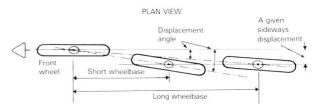

Fig. 2.15 **A long wheelbase enhances directional stability by reducing the displacement angle of the rear wheel**

Summary
In common with most design features, wheelbase is a compromise and varies with the intended use of the machine. Trials machines need good manoeuvrability, mainly at low speed, and so have short wheelbases. Large touring machines need good directional stability for relaxed riding but not the razor-sharp handling of a racer; hence touring machines have longer wheelbases (about 58–60 in.), though too long a wheelbase impairs manoeuvrability in traffic. Some feet-first machines, such as the out-of-production Quasar, have wheelbases of 77 in. or so. A racer must compromise between the requirements of stability at very high speeds, quick handling and minimum weight transfer. Actual figures tend to be around 50 in. in the smaller classes and 56 in. for the larger, faster machines.

Wheel diameter
In road racing, experiments are being carried out to determine whether 16 in.-diameter wheels are preferable overall to the 18 in. size previously used almost exclusively. As yet, there does not seem to be a definite concensus. This is yet another area for compromise, as there are advantages and disadvantages on both hands, as follows:

1) For a given tyre section, a small wheel reduces both the unsprung mass (to the bene-

Above **Long wheelbase. Because of the riding position, the wheelbase on the out-of-production Quasar was of necessity some 20 in. longer than normal for a touring machine, thus enhancing directional stability at the cost of manoeuvrability (Dixon)**

fit of roadholding) and the steering inertia. This is welcome in all cases.

2) Wheel size also effects gyroscopic forces. For a given tyre and rim section, these forces are proportional to the road speed and the square of the wheel diameter. Thus, bigger wheels will start to give their balancing effort at lower speeds.

3) As figure 2.16 shows, a smaller wheel drops farther into holes; similarly, it feels raised bumps more sharply. So touring machines need larger wheels (for comfort and roadholding on rough roads) while trials and motocross machines have 21 in. front wheels, the better to ride the bumps. True, these bikes have 18 in. rear wheels but the large tyre section there

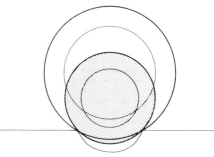

Fig. 2.16 **A smaller wheel drops farther into holes than a larger wheel**

brings the overall diameter up to that of the front.

4) For a given tyre section, the area of rubber on the ground is greater with larger wheels. With smaller wheels we could restore the area by widening the tyre (as is done on 16 in. racing wheels) but this might bring other problems.

5) The self-steering effect of trail and rake

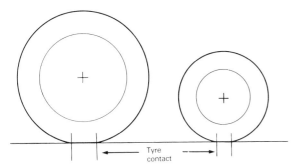

Fig. 2.17 **For the same section, a larger-diameter tyre gives more contact area on the road**

mentioned earlier is emphasized by the use of smaller wheels.

6) For purely structural reasons, smaller wheels are stiffer laterally.

To sum up on steering geometry, perhaps it is time to rethink our ideas on the relationship between rake and trail and their effects on handling and stability. Experiments are needed to establish the practical effects of steeper steering axes, although some hub-centre layouts already provide a pointer with their 5–15-degree angles. It may well be that the best results will come from rake angles between zero (where the adverse effects of rake are non-existent) and the angle that gives the required trail with zero offset, for minimum steering inertia. See page 58.

Castor instability

While trail provides a degree of stability, it is ironic that it can itself cause an oscillating, or wobbling, type of instability. This happens if, when the front wheel is displaced by some road irregularity, the restoring force created by the trail happens to be strong enough to over-correct for the initial disturbance. The wheel will then swing beyond the straight-ahead position and will be steering in the opposite direction. This in turn creates another restoring

force, which repeats the whole process, and we have a side-to-side steering wobble caused by the trail.

The above is a simplified explanation of a phenomenon known as simple harmonic motion; the essential elements are a mass free to move but restrained by a form of spring or other means which generates a force (dependent on the displacement) tending to restore the mass to its rest position or beyond. The best example of this that has some similarity to our steering problem is the pendulum. Consider figure 2.18.

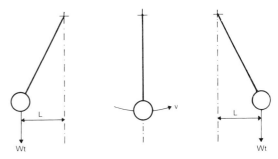

Fig. 2.18 Left: **With the pendulum in this position the weight (Wt) acts on the lever arm (L) to accelerate the mass towards the centre.** *Middle:* **In this position the pendulum is moving with velocity (v) sufficient to carry it beyond the centre to the other side.** *Right:* **When the pendulum reaches this position all its kinetic energy is given up in favour of potential energy and the cycle is repeated**

As we move the pendulum to one side, gravity creates a force that starts to accelerate the weight back towards the centre; but, by the time it gets there, it is moving fast enough to carry on to the other side and the whole process continues, with the pendulum swinging from side to side.

If we wait long enough, the extent of the swings will progressively diminish until the pendulum eventually comes to rest. This is due to damping, which may take many forms. In the pendulum's case, the damping comes mainly from air resistance and friction in the pivot. The amount of damping determines the number of swings before the pendulum comes to rest.

It is possible, however, to have an oscillating system that doesn't come to rest in this way or which actually increases the magnitude of successive swings (sometimes until the system is destroyed). For this to occur we need negative damping, which may also be termed a forcing function. This simply means the application of some added impetus to the system at the appropriate time. In the case of a pendulum in a clock, the impetus comes from a spring (or a weight on a chain) which, through the escapement mechanism, gives the pendulum a kick on each swing to ensure that it does not come to rest.

A child's garden swing is just a special case of pendulum. We have only to give it a gentle push each time it reaches a peak for the amplitude of the swing to increase very quickly; in extreme cases the swing may even be made to go over the top. If the gentle push is timed wrongly, however, it is surprising how quickly the swing can be brought to rest.

(We make no apology for this lengthy discussion of swings and pendulums because a basic knowledge of simple harmonic motion is essential to an understanding of the causes of several motorcycle problems and of the function of suspension systems to be described later.)

Now let us examine the concept of resonance. Experiments with our pendulum or swing will soon show that (except for extreme angles of swing) the number of complete oscillations in a given time is almost constant, regardless of the amplitude of the swing. This is called the resonant frequency (or natural frequency) and is mainly determined by the length of the pendulum.

As we have seen, if our push (forcing function) to the child's swing is correctly timed (i.e.

in phase) and at the natural frequency it is easy to build up high amplitudes of oscillation. But if the forcing function is either out of phase or of a different frequency, then the oscillation will be much less inclined to build up and may even be heavily supressed.

If we apply the principles of simple harmonic motion to a motorcycle we find that our wobbling front wheel has a natural frequency determined mainly by: 1) the moment of inertial of the front wheel and fork about the steering axis; 2) the magnitude of the restoring torque due to a given angular displacement; this is determined by the rake, trail, tyre size and the characteristics and stiffness of the frame and fork.

The *higher* the steered moment of inertia, the *lower* the natural frequency; and the *higher* the restoring torque per degree of deflection, the *higher* the natural frequency.

Other wobbling systems are possible on a motorcycle and their interactions affect the overall characteristics. With most heavyweight touring machines, for example, if we stand alongside and shake the handlebar rapidly from side to side, then the rear of the machine will develop a surprising sideways oscillation, usually more pronounced if panniers and luggage are fitted. Close examination will reveal that most of the movement takes place in lateral flexure of the rear tyre. This oscillation has its own natural frequency, affected by passenger and luggage weight and tyre properties. If this (or other) resonance coincides closely with the front-wheel wobble, then the overall effect may be greatly intensified.

The forcing function trying to build up a wobble on a motorcycle results from unavoidable imperfections in the machine. A small unevenness in a tyre or a slight buckle in a rim will give a kick to the steering at each revolution of the wheel, as will any wheel unbalance that is offset from its centre plane. The frequency of these forcing functions is that of the rpm of the wheel (or a multiple of the rpm); and if our speed is such that this coincides with the natural wobble frequency, then the bike may develop a bad wobble at this speed.

Several factors contribute to the fact that all bikes don't wobble in this way—and damping is an important one. Not only the deliberate damping provided by a steering damper but also damping inherent in the wobbling system, such as friction in the steering bearings, wiring and control cables, tyre friction and internal tyre damping (hysteresis); often the decisive factor is damping by the rider's body through his contact with the handlebar. Ultimately, the natural wobble frequency, road-induced forcing function and damping fight it out together and the result determines the machine's wobble characteristics.

Some machines are free from this type of wobble, yet on others it is difficult to eliminate. It usually occures between 25 and 40 mph and is felt most strongly, sometimes violently, while slowing through this range with the hands removed from the bar. As an example, BMWs with the Earles-type pivoted front fork (pre-1970) could develop a pronounced wobble at about 35 mph, especially when carrying a pillion passenger or heavy luggage at the rear. An hydraulic steering damper effected a cure without any adverse effect.

The most important point to bear in mind is that the fundamental mechanism for causing this sort of wobble is inherent in the layout of a conventional motorcycle; the only way to prevent it is to damp or tune it out of the system and the following measures may be helpful:

1) Increase frame and fork stiffness; this raises the natural frequency of the wobble (see also 3, below).

2) Reduce the trail; there are limits to this approach as it may impair directional stability.

3) Reduce the weight of the front wheel and fork, so reducing their moment of inertia about the steering axis. This cuts down the energy in the oscillating parts for a given magnitude and frequency of wobble, so that the inherent damping has a greater proportional effect. This reduction of inertia also raises the wobble's natural frequency, so that the gyroscopic forces (which become very strong at high speeds) are better able to resist the tendency to wobble. However, these same gyroscopic forces will also tend to convert any steering movement into a leaning movement; this further complicates the issue, for we now have a flopping motorcycle as well as a wobbling front wheel. This cross-coupling between steering and banking effects is accentuated by the geometric effects of rake and trail and is one reason for the influence of a passenger and rear luggage on wobble characteristics.

4) Fit a hydraulic steering damper.

Some other considerations

1) Frame stiffness

In building a chassis suitable for our purpose, we have more problems than those involved in simply achieving the best compromise between the various geometric parameters discussed so far. For if the chassis is not rigid enough to maintain our chosen geometry in use, then all our calculations will be set at nought.

There are many sources of flexure in a motorcycle and all must be dealt with to achieve good handling. It is especially important to maintain the alignment between the centre planes of the wheels and the steering axis, otherwise directional stability will suffer and the tendency to wobble may be increased.

At the front end (which is the more important) this alignment is governed mainly by the lateral stiffness of the fork and wheel. Hub-centre steering scores heavily here because, in most layouts, only the wheel constitutes a possible source of flexure.

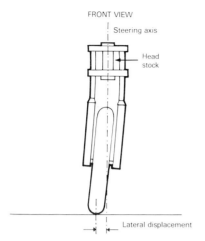

Fig. 2.19 **Lateral displacement due to fork and wheel flexure**

Telescopic forks supported by a conventional high steering head usually lack stiffness in a fore-and-aft plane; but this is of little concern except perhaps under braking, when the weaker designs may give rise to shuddering and wheel hop. This is yet another problem reduced by hub-centre steering.

A practical point that is sometimes overlooked in connection with handling is the lateral stiffness of the sub-frame supporting the seat. The rider receives much of his feedback on the machine's behaviour through the

Fig. 2.21 **The long lever arm causes high bending moments in the fork legs and steering head; this can give rise to juddering and wheel hop**

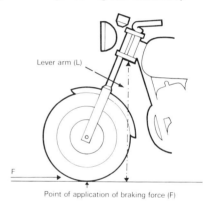

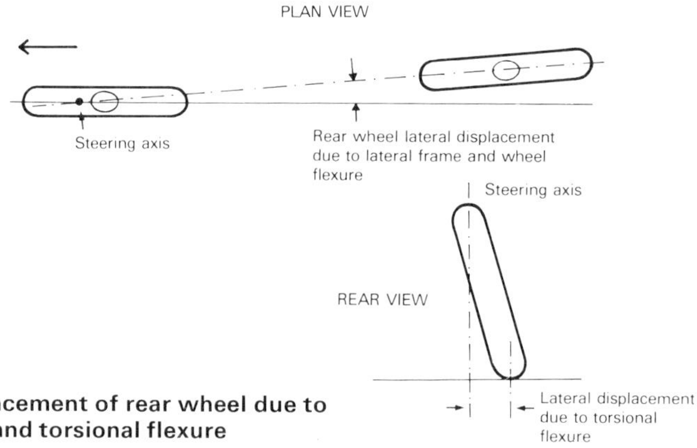

Keeping the rear wheel aligned with the steering axis involves not only the lateral stiffness of the wheel but also the torsional and lateral stiffness of the main frame and pivoted rear suspension.

In normal riding, torsional stiffness between the handlebar and the front-wheel spindle is not normally a problem. But it is very important in trials and motocross, where high steering torques are often necessary to get out of ruts.

Fig. 2.20 **Lateral displacement of rear wheel due to lateral frame flexure and torsional flexure**

proverbial seat of his pants; if the seat itself is flapping about independently of the chassis he will get the wrong message. Many cases of suspected bad handling have been rectified by stiffening the seat support. Handling of the featherbed Manx Norton became noticeably more taut when the rear subframe was welded, instead of bolted, to the main frame; this modification also cured a spate of fractures of suspension-strut piston rods, probably caused by lateral flexure.

2) Weight (mass) and its position—inertia effects

Generally speaking, the less mass a machine possesses the better. Under the influence of a given force, the smaller the mass the more quickly it will accelerate. Not only does this mean a brisker performance for a given engine

power, it also means a more responsive handling performance for a given effort by the rider.

Just as important as the amount of a machine's mass are its distribution and the location of the mass centre, as the following considerations show.

Balance Low weight and a low centre of gravity both facilitate good balance. Figure 2.22 shows how, for a given degree of lean, the unbalancing couple is directly proportional to the weight and the height of the centre of gravity.

Angle of lean The angle of lean necessary to balance centrifugal force when cornering is slightly affected by the centre-of-gravity height. See Figures 2.23 and 2.31.

Weight transfer Under braking, weight is transferred from the rear wheel to the front; under acceleration, the transfer is in the opposite direction.

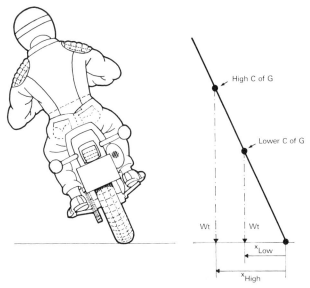

Fig. 2.22 The unbalancing couple is equal to **Wt.** × **x** i.e. the weight (**Wt**) multiplied by the lever arm (**x**). Since the lever arm is proportional to the centre of gravity height, a higher centre of gravity gives a higher unbalancing effect

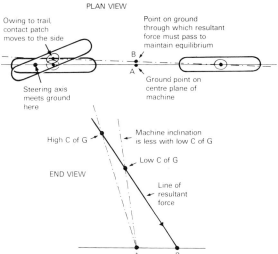

Fig. 2.23. **Although a lower centre of gravity requires a smaller banking angle, the effect is very small in practice and countered by an opposing effect due to the finite width of the tyres (see Fig. 2.31)**

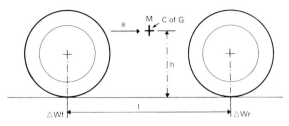

Fig. 2.24 **Where**

$$l = \text{wheelbase}$$
$$h = \text{c of g height}$$
$$M = \text{machine mass}$$
$$a = \text{braking acceleration}$$
$$\Delta W_f = \text{weight transfer at front}$$
$$\Delta W_r = \text{weight transfer at rear}$$

and let g = **gravitational constant**

Then

the machine at the C of G of $M\frac{a}{g}$. This force acts

through the lever arm h to form a couple $M.\frac{a}{g}.h$.

This couple is resisted by an opposite couple due to the weight transfer acting on the length l.

$$\text{Therefore } \triangle W_f l = \frac{Mah}{g}$$

$$\text{and } \triangle W_r l = -\frac{Mah}{g}$$

$$\text{thus } \triangle W_f = \frac{Mah}{l}.\frac{a}{g}$$

$$\text{and } \triangle W_r = -\frac{Mah}{l}.\frac{a}{g}$$

We can thus see that weight transfer is proportional to the bike's mass, c of g height and acceleration, and inversely proportional to the wheelbase

Lengthening the wheelbase decreases the weight transfer, as also does lowering the mass centre height and reducing the mass. Weight *transfer* is not affected by the longitudinal location of the centre of gravity, though this obviously controls the static weight supported by each wheel.

Traction Since the driving force the rear wheel can deliver to the ground is limited by the load carried by the wheel, a rearward

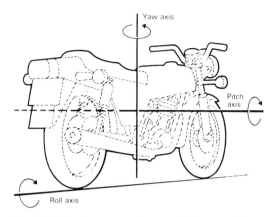

Fig. 2.25 **The three axes about which a machine can rotate**

weight distribution improves traction. However, we must balance this requirement against the need to keep the front wheel on the ground for steering. A forward mass bias also helps directional stability, as it does in a dart or an arrow.

Angular motions So far as linear motions are concerned, it is the *amount* of the machine's mass that is important. But when it comes to the angular motions of pitch (about a transverse axis), yaw (about a vertical axis) and roll (about a longitudinal axis), the *distribution* of the mass is all-important because that governs what is called the moment of inertia.

This is a measure of the inertia effect about the particular axis and its value determines the ease with which we can accelerate the machine about that axis.

(a) *Roll:* The roll axis is obviously the line joining the centres of the two tyre contact patches. And the roll moment of inertia is the sum of all the individual components of the total mass multiplied by the square of their distance from the roll axis. (The moment of inertia is sometimes said to be the product of the total mass and the square of the distance to the centre of gravity; this is at best only an approximation.)

A low roll moment of inertia is desirable for a rapid and effortless change in banking angle (say, through an S-bend) and this usually means a low centre of gravity.

(b) *Pitch:* It is not so easy to define the axis about which a machine pitches because it varies with the bike's configuration. For example, if a machine is sprung at the front but not at the rear, it will pitch about the rear-wheel centre point, while a machine with the opposite arrangement (i.e. sprung rear, rigid front) would pitch about the front-wheel centre.

In the case of a conventional machine, sprung at both ends, the pitch axis depends on suspension geometry and spring rates. Except in trials and motocross, there is no great need for a fast pitch response and so a large amount of inertia is not harmful. Indeed, it may contribute to comfort when braking or traversing bumpy surfaces.

The tail fin on a fully streamlined high-speed record machine enhances directional stability by moving the centre of pressure rearward. Here H. P. (Happy) Mueller streaks across the Utah salt at 150 mph in the 125 cc NSU flying hammock in 1956 (*MCW*)

(c) *Yaw:* In this case there are conflicting requirements for both a high moment of inertia and a low one. For example, a high value enhances directional stability while a low value facilitates rapid changes of direction and minimizes the effects of a slide.

Within practical limits, it is found better to aim for a low moment of inertia, which involves concentrating the mass of the bike as close as possible to the longitudinal centre. Naturally this tends to a low pitch moment of inertia too.

3) Aerodynamic effects

The size and shape of the machine—together with the rider and any aerodynamic aids such as a fairing, windscreen or legshields—affects its drag and lift, hence its power requirement, the more so as speed increases. To a large extent, this aspect of design is outside the scope of this book; but it is worth noting that the behaviour of the air as it leaves the machine at the rear is usually more important than what happens at the front.

Indeed, machines designed for very high-speed records nowadays not only have the smallest possible frontal area but streamlining extended well behind the rear wheel, with a large vertical stabilizing fin. Fairing design for roadsters and road racers usually leaves much to be desired, for most designers concentrate their attention on the front end and neglect the rear. In racing, this is a result of the tight framework of FIM rules to which fairings must conform. In touring, it is due to designers either copying racing designs or thinking only of weather protection for the rider.

In much the same way as a weather vane keeps its arrow pointing into wind, so the tail fin on a high-speed record machine gives it directional stability.

Let us see how this works in practice. Suppose, for whatever reason, the machine adopts a yaw attitude at high speed, when aerodynamic forces are significant (i.e. it is point-

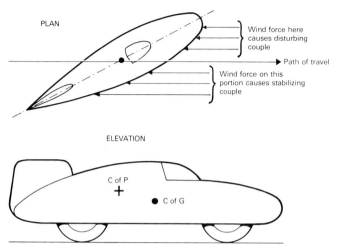

Area of large tail fin ensures a rearward centre of pressure (C of P) for directional stability

Fig. 2.26 **Stabilizing effect of tail fin**

ing at a slight angle to the direction of travel, as shown in the sketch). Initially, the machine's momentum tends to keep it moving in the original direction, regardless of its attitude, so that the wind force created by its motion acts on the whole of the forward-facing flank.

The pressure on the area ahead of the mass centre gives rise to a moment that reinforces the disturbance, while the pressure on the area behind the mass centre tends to correct the yaw. Which of these opposing effects predominates depends on the relative positions of the mass centre and the centre of pressure. For, just as the mass centre is the point through which the inertia forces act on the machine, so the centre of pressure is the point where the aerodynamic side forces can be said to act. For aerodynamic stability, the centre of pressure must clearly be behind the centre of gravity—and it is the additional side area of the tail fin that ensures this.

Side winds The effects of these are best considered under two separate headings: steady state and dynamic.

(a) Steady state means riding along with a uniform side wind, which thus creates a constant side force acting through the centre of pressure. Clearly, this causes a couple about the roll axis, which must be balanced by leaning into the wind. To minimize the angle of lean required to balance a given side force, we need a low centre of pressure and a high centre of gravity and weight. Figure 2.27.

To prevent the side wind from gradually blowing the bike off course, it must be steered into wind to compensate. This occurs automatically if the centre of pressure is the correct distance behind the 'centre of tyre grip' (which is practically the same as the longitudinal centre of gravity). If this is not the case, then the rider must apply a correcting torque to the steering.

(b) The dynamic state refers to riding in gusty conditions, such as when passing gaps between hedges or buildings or when overtaking large pantechnicons.

As in the steady state, we have to steer into the wind to maintain the required direction; again this is automatic provided the centre of

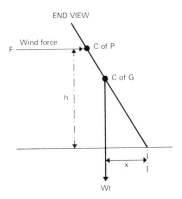

Fig. 2.27 To balance moments, Wt.x = F.h
Therefore to reduce the angle of lean we need a high value of Wt.x, i.e. a high weight value and c of g position. NB: for moderate angles, x is proportional to the angle of lean

pressure is behind the mass centre. The rolling couple (from the wind) is balanced by accelerating the roll moment of inertia about its axis; but any quick roll motions produce steering effects (through gyroscopic precession) which will oppose our need to steer into the wind. So it is important to keep the banking change to a minimum, which means a low centre of pressure and a high roll moment of inertia.

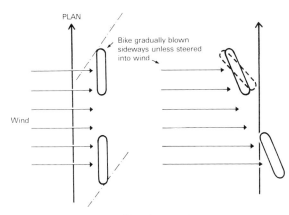

Fig. 2.28 Side wind effect in steady state

Anti-lift Since racing cars have achieved dramatic increases in cornering speed through the extra tyre grip produced by aerodynamic downforce from front and rear wings and 'ground-effect' chassis, it might be thought that the same principle could be exploited on a solo motorcycle. Because our machine has to be banked for cornering, however, the situation is more complex than it is on a car. Indeed, a fixed wing or other means of generating aerodynamic downthrust might well reduce our cornering speed. Figure 2.29 representing

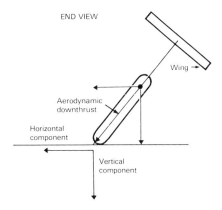

Fig. 2.29 A fixed wing increases the horizontal and vertical components of the tyre force in the same proportion

a banked motorcycle, shows that the downforce provided by a fixed wing acts in line with the bike and has both vertical and horizontal components. But although the vertical component is available at the tyre contact patch to increase grip, this extra grip only counteracts the horizontal component.

As explained in the next section, available tyre grip depends also on the contact-patch pressure, with the result that the extra tyre loading from the downforce reduces the coefficient of friction. Thus the grip available to resist the centrifugal cornering force is also reduced, so that cornering speed is slower.

Conversely, aerodynamic *lift* might actually increase cornering speed—though at the expense of high-speed braking and traction forces and of stability. Taken to the extreme, a motorcycle would then become an aeroplane, which can corner at lateral accelerations limited only by the ability of the structure to withstand the forces generated.

If we could generate aerodynamic downthrust in a vertical plane only (say, by a tilting wing) then increases in braking, traction and cornering speed would follow (as in a car). An interesting side effect would be that a smaller angle of lean would then be required for a given cornering force. As can be seen from figure 2.30 the moment due to the downforce counteracts the centrifugal moment. Hence the moment due to the machine's weight must be reduced—which means a smaller angle of lean. However, any benefits from such a device might well be outweighted by practical and stability problems.

4) Tyres

To explore carcass construction, tread compound and tread pattern in detail is beyond the scope of this book. But rather we are concerned here with some basic principles and their effects on handling characteristics.

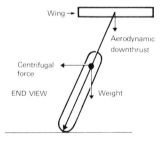

Fig. 2.30 **In theory, a tilting wing that remained horizontal could enhance braking, traction and cornering power. As a side effect, a smaller angle of lean would be required for a given cornering speed**

Since they constitute our only contact with the ground, the tyres are crucial in providing the grip to transmit driving, braking and cornering forces. The amount of grip depends on the weight supported by each tyre; increasing the weight increases the grip.

The ratio between the maximum possible grip and the load is called the coefficient of friction. However, this coefficient is not constant but usually decreases with load (i.e. increased contact-patch pressure). A further complication is that this relationship is not linear.

This has far-reaching implications and is one reason for the general increase in tyre section because, for a given wheel load, the bigger the section the lower the contact-patch pressure and so the greater the coefficient of friction, hence the grip on the road.

Heavy braking (when as much as 90 per cent of the total weight may be supported on the front tyre) provides an interesting example of the effects of the relationship between load and friction coefficient. The forward weight transfer increases the pressure on the front contact patch and reduces that on the rear, so reducing the coefficient at the front and increasing it at the rear.

So the tyre with the reduced coefficient of friction is carrying most of the weight and vice versa. Hence the total frictional force available for stopping is less than it would be on a machine with a smaller weight transfer. In other words, for maximum braking we need a long wheelbase and a low mass centre.

Another reason for the trend towards larger tyre sections is, of course, the relentless growth in the weight and power of our machines, which would otherwise cause excessive tread wear.

To balance the grip forces with the individual loads, different tyre sections are used on the front and rear wheels. Since a motorcycle with the rider on board usually has a rearward weight bias, a larger tyre section is used at the rear.

If we overtyre at the rear (or undertyre at the front) then the coefficient of friction at the front will be less than that at the rear, giving rise to understeer. Alternatively, if we overtyre at the front (or undertyre at the rear) then the effect will be oversteer. In either case, the end with the smaller coefficient of friction will lose adhesion first, at a lower cornering speed than if that tyre section was increased to balance. Compromise is unavoidable here because a change in throttle opening or in wheel loading (through carrying a passenger or heavy luggage) can drastically alter the front/rear requirement.

The profiles and lateral flexing characteristics need careful balancing front to rear. Indeed, both the tyre and bike manufacturers undertake considerable testing to determine the best blend; a combination of tyres that gives excellent results on one model may prove near lethal on another.

Relative to their weight and power, motorcycles have smaller tyre sections than cars. This derives chiefly from our need to bank while cornering, in which case wide tyres may impair handling.

Figure 2.31 shows how the contact patch moves away from the centre plane of the wheel or the steering axis as the machine is banked, giving rise to the following side effects:

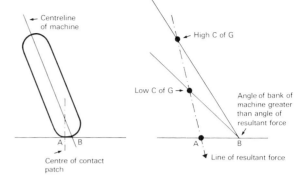

Fig. 2.31 **This shows the different angles of lean required with low and high centres of gravity, due to the width of the tyre. The effect opposes that shown in Fig. 2.23**

a) A greater angle of lean is necessary to balance centrifugal force; this may involve a higher centre of gravity to restore cornering clearance.

b) Because the forces fed into the tyre are offset from the steering axis, steering forces are introduced which have to be counteracted by the rider, usually requiring more effort.

The adverse effects of large tyres are not confined to cornering. Even in straight-ahead riding, they may cause problems.

As figure 2.32 shows, a road disturbance such as a stone can cause a couple tending to steer the wheel to that side.

Large tyres increase the unsprung mass too, to the detriment of roadholding and comfort,

Fig. 2.32 **Steering torque produced by road disturbance**

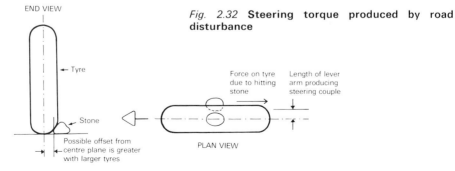

though the extra suppleness of fatter tyres makes some contribution in these areas.

Larger tyres also increase precessional forces.

Finally, a fatter section magnifies a tyre's self-steering characteristic. This is the rolling-cone effect and is due to the fact that the inner edge of the contact patch is at a smaller radius from the wheel spindle than is the outer edge (see figure 2.33).

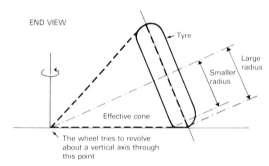

Fig. 2.33 **Self-steering effect of a banked tyre (rolling-cone effect)**

5) Braking stability

Our examination of braking in the previous section showed how the coefficient of friction of the front and rear tyres varies with forward weight transfer—and how overall braking efficiency benefits from keeping that transfer to a minimum. The requirements for directional stability under heavy braking are more involved and need careful thought.

As with our earlier consideration of castor effects, there is a basic lack of stability inherent in braking too, and this must be countered by the rider if he is to maintain control.

Suppose a bike, while braking, is deflected from its true direction of travel, as in figure 2.34. The inertia force (F_I) gives rise to an unsettling couple ($F_I.x_1$) about the front tyre contact patch. At the same time, the rear braking force (F_R) creates a correcting torque

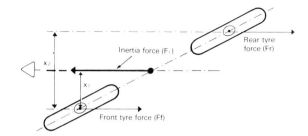

Fig. 2.34 **Yaw moments under braking. To balance the moments about the front wheel, the stabilizing moment from the rear brake force ($Fr.x_2$) must equal the destabilizing couple from the inertia force ($F_I.x_1$)—i.e. $Fr.x_2 = F_I.x_1$**

($F_R.x_2$). For natural stability the correcting torque must exceed the unbalancing one.

This occurs only under low or moderate deceleration, when there is little weight transfer and so the rear wheel can provide a relatively high proportion of the total braking. Heavy braking, on the other hand, involves considerable forward weight transfer, so that the front wheel is required to do the lion's share of the work, and the unsettling couple then exceeds the correcting torque—an unstable condition.

If we are to reduce this tendency we need to minimize the weight transfer by means of a long wheelbase and low mass centre. A rearward centre of gravity also helps by permitting more braking at the rear to increase the realigning couple; but this conflicts with the need for a forward mass centre to reduce the destabilizing torque.

To complete this analysis let us consider four separate braking conditions.

Rear brake only The rear tyre force balances the inertia force and the resulting couple tends to stabilize the machine.

Front brake only Here the above situation is reversed so that the couple tends to destabilize the machine.

Both brakes applied, rear wheel locked As soon as the wheel locks, its tyre loses most of its grip and this can have two effects: first, we lose most of the rear brake's stabilizing effect; second, because the total braking force is reduced there is less forward weight transfer and, if we are already braking close to the limit, this may cause the front wheel to lock, too, with disastrous results.

Both brakes applied, front wheel locked Here we have strong directional stability from the still-braking rear wheel; also the reduction in weight transfer restores some weight to the rear, so permitting increased braking at that end.

All these cases refer only to directional stability while braking in a straight line. The situation changes drastically when negotiating even a mild bend while braking. As soon as either wheel locks then, it loses most of its ability to provide much sideways force to balance the centrifugal load of cornering, and the wheel slides outward. Depending on the

Diagrammatic layout of Brembo hydraulic coupled brakes. The object is to eliminate reliance on a high degree of expertise in apportioning front and rear effort (Brembo)

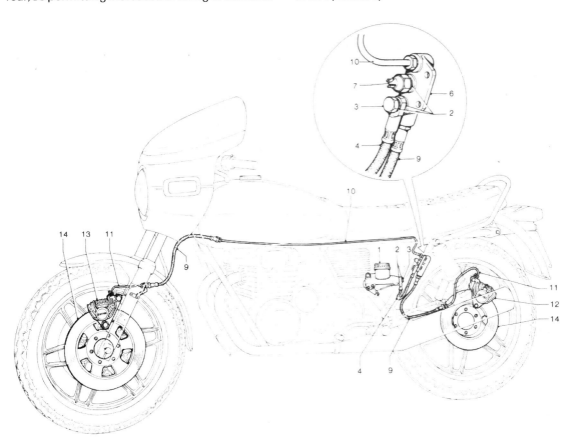

rider's skill (or luck) it is usually much easier to recover from a rear-wheel slide than a front one.

As explained earlier, the balance of a single-track vehicle depends on the precessional forces of rotating wheels, especially the front one—hence the desirability of not locking the wheels. The situation of a locked front wheel and a braking rear wheel, though directionally stable, is unstable so far as balance is concerned.

Unlike cars, where all four brakes are controlled by a single pedal and the front/rear balance is pre-determined by the manufacturer, most motorcycles have separate controls for the front and rear brakes in the questionable belief that the rider himself is best able to judge the required balance in all circumstances. This may be so for a highly skilled rider but is unlikely for the less expert.

Moto Guzzi, for one, have introduced coupled braking in an attempt to reduce the degree of expertise required. In their system the pedal controls the rear brake and one of the two front brakes, while the other front brake is controlled by the handlebar lever. Thus additional front braking is available if required.

Since the degree of forward weight transfer depends on how hard the brakes are applied, it is clear that optimum braking calls for a greater front/rear bias in dry conditions than wet.

Because the consequences of a locked wheel can be severe, it is to be hoped that microprocessor-controlled anti-lock will soon be commonplace. The technology has existed for some time—its application would make a tremendous contribution to safety.

Appendix to Chapter 2

Experiments with rake and trail variations

1) Rake The ideas expounded in the previous chapter in relation to the angle of the steering

axis (rake) were subsequently put to the test by modifying a readily available standard production machine—a BMW R75/5.

There were two technical advantages in the choice of this machine. First, the offset of the wheel spindle from the steering axis is divided almost equally between the offset in the yokes and that of the wheel spindle from the centre line of the fork sliders (figure A2.1); the

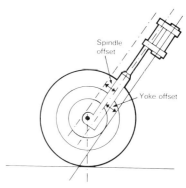

Fig. A2.1 **In the BMW R75/5, the total offset (wheel spindle from steering axis) is divided approximately equally between that in the yokes (steering axis to fork legs) and that in the sliders (fork legs to spindle)**

importance of this will become obvious later. Second, the BMW was large and fast enough to make the results meaningful, which might have been less so with a slow, light machine such as a moped.

To keep other variables to a minimum, the original frame and suspension were retained and the wheelbase remained unaltered.

Two non-standard rake angles were tried. In each case the trail was kept to approximately the same as the standard value (i.e. $3\frac{1}{2}$ in.).

The first alternative set-up tried was with a rake of approximately 15 degrees and almost nil offset. This was achieved by bolting a superstructure to the frame to support the new headstock (see photo) and reversing the yokes; since their offset is very close to that

Trail

The standard BMW R75/5 used as a basis for the experiments in rake and trail; maker's figures were 27 degrees and $3\frac{1}{2}$ in.

of the wheel spindle, the overall offset was reduced virtually to zero.

For the second set-up the rake was close to zero (i.e., vertical steering axis). This was achieved by reversing the complete front-fork assembly, thus giving the negative offset necessary to maintain the standard trail. The new headstock was supported by an extension of the original superstructure.

In both cases the handlebar was pivoted in the usual place and connected to the fork by a ball-jointed link; a side effect of this was an adjustable steering ratio—i.e., for a given angle

at the fork the angle needed at the bar could be varied.

With the 15-degree rake the bike had full road equipment, including lighting, so that it could be ridden under everyday conditions; indeed, five riders covered nearly 2000 miles between them, including wet and dry going, bumpy country lanes, London traffic and motorway trips. Throughout this period no steering damper was fitted.

Although the results of these tests are essentially subjective and might be expected to depend on experience, personal preferences and preconceived ideas, there was in fact no divergence of opinion.

The initial testing was done on a bumpy, rutted country lane at speeds up to 50 mph. Here

Trail

Trail

the most noticeable effect was the total insensitivity of the steering to ruts and bumps. Not only could the bike be ridden hands-off but at the same time it could be weaved from side to side across the ruts with little effort and no detectable deflection of the steering.

In corners, bumps had little effect, which was contrary to the behaviour of this particular machine before conversion, when it had a strong tendency, with no steering damper, to shake its head (sometimes violently) on bumpy corners.

This lack of disturbance by longitudinal ruts was also confirmed on smoother roads at higher speeds, when the machine was ridden deliberately on the edge of painted white lines. Though unforseen, this benefit is easily explained by reference to figure A2.2.

If we visualize a 90-degree rake (i.e. horizontal steering axis) we can see that the side of the rut gives rise to a moment about the steering axis that tends to steer the wheel back into the rut. With a vertical steering axis, however (zero rake), there is no effect on the steering; instead, the disturbance tends to cause the complete machine to lean into the rut. In this case, though, since the inertia of the whole bike is much higher than that of the front wheel alone, the effect on directional stability is considerably smaller and the rider is less aware of the rut. Thus the steeper the steering axis the smaller the effect.

In the foregoing chapter, we suggested that balance might be enhanced, particularly at low

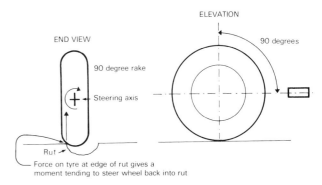

Fig. A2.2 **The effect of ruts on steering increases with rake angle, as shown in this exaggerated case. A vertical steering axis reduces the effect**

speeds, by steepening the head angle. To verify this, much riding was done at very low speeds and balance was indeed improved by the modifications. The machine could easily be ridden much more slowly than when in standard trim before the rider had to put a foot down. (Of course, champion trials riders can balance a stationary machine indefinitely; but that is exceptional and most of us need to be moving slightly to maintain balance.)

In heavy traffic, it was noticeably easier to trickle along slowly on the modified BMW, making it less tiring to ride from one side of London to the other. When, without prior briefing, a novice was asked to try the machine, he commented on the surprising ease of moving off from rest; there was less wobbling than usually seen with a learner and his feet were quickly on the rests.

It has been suggested that an unusually steep steering axis might induce wobbles at high speed. But with both the experimental rake angles on the BMW (15 degrees and zero) this was not noticed. With the handlebar released, the machine was ridden from approximately 100 mph down to a walking pace and

Left, top **In the first experiment a superstructure bolted to the frame steepened the steering axis to 15 degrees while leaving the trail unaltered. Note zero offset**

Left, bottom **By extending the superstructure and reversing the complete fork assembly, rake was reduced to near zero for the second experiment, again with no alteration in trail. Note negative offset**

With the 15-degree rake, Tony Foale demonstrates the stability of the modified BMW

at no time was there any tendency to wobble or weave.

With confidence built by several such runs, the handlebar was knocked to try to initiate a wobble. Whatever the speed, though, the disturbance was damped out within less than one cycle. In standard trim (27-degree rake) this particular machine could develop a pronounced wobble when ridden no-hands at 30 to 40 mph, though it was easily damped out by grasping the handlebar.

Directional stability was always excellent and tremendous confidence was instilled in the rider at an early stage.

A further advantage of the steeper head angles was increased sensitivity of the front fork to small bumps. This results from reduced 'stiction' in the fork sliders as a consequence of the decrease in side loading. (The normal side-load component is approximately halved by reducing rake to 15 degrees and practically eliminated at zero rake.)

Also, this reduction in the side-load component is accompanied by an increase in the spring-load component as the fork is steepened—which gives the same effect as a lower spring rate. The effective rate varies little between zero and 15 degrees rake but is approximately ten per cent higher at 27 degrees.

Similarly, the spring-load component of the braking force is reduced as the fork angle is

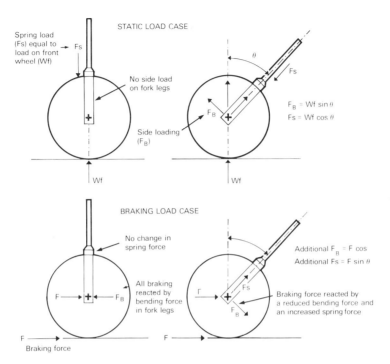

Fig. A2.3 **Steepening the steering head reduces the stiction in telescopic fork sliders, so improving sensitivity to small bumps. Nose diving under braking was also reduced**

steepened. And since this spring force acts in concert with weight transfer to compress the fork, the reduction means less nose diving. For this reason the effective drop in spring rate was not detrimental and ride comfort was appreciably improved.

It was under braking, however, that a disadvantage was noticed, in the form of severe shuddering in the fork as the braking force tried to bend it backward. Naturally, this was more severe with the steering axis upright, even though the reversal of the forks converted the twin-leading-shoe brake to much less effec-tive twin trailing shoes. Such juddering was entirely absent with the standard rake, though quite bad at 15 degrees.

This effect apart, one of the most interesting results (mentioned by all the riders) was the surprisingly normal feel of the modified machine, with the steering pleasantly light at low speeds but always totally stable. No special riding technique was required and cornering was accomplished normally.

The variable steering ratio mentioned earlier was tried from 1:1 (equivalent to conventional direct steering) to 1:2 (steering angle doubled from handlebar to fork). In normal riding (dry roads) it was impossible to detect the difference; only when manoeuvring at a standstill, using large steering angles, was the heavier feel of the 1:2 ratio noticeable.

Rake angle

Trail

Above **In both the experiments the handlebar was connected to the fork by a Rose-jointed link providing for steering ratios anywhere between 1:1 (direct) and 2:1 (geared up)**

However, since the steeper rakes made the steering lighter anyway, even the effort required with the 1:2 ratio was similar to that with the standard machine.

Indeed, the reduced handlebar swing could be a bonus when designing a non-steerable fairing, as handlebar clearance usually results in a bulky shape if steering lock is not to be restricted.

2) Trail Throughout the rake experiments the standard trail of approximately 3·5 inches was retained. Yet it seemed reasonable to assume that the optimum trail (if there is such a thing) might vary with rake angle.

To test this, a double-wishbone front-suspension system was fitted to the BMW (see

To try trail values between 2 and 4 in. (with a rake in the region of 15 degrees) this adjustable double-wishbone suspension was made up

photo), making it possible to alter the trail by adjusting the wishbone lengths to give variations in rake. There was no offset, and trail values from 2 in. to 4 in. were tried varying the rake angle either side of 15 degrees.

Although the machine was perfectly rideable over the full range of settings, the front end proved livlier and the steering more sensitive to bumps as the trail was shortened (steeper rake), albeit not so much as with the standard set-up. In the upper range of trail settings the bike was very steady but could still be manoeuvred quickly.

At about 3 in. trail there was a tendency for the machine to lift itself out of a bend when cornered at moderate banking angles (say, 15 to 20 degrees), though no such effect was detectable at higher cornering speeds requiring, say, 35 to 45 degrees of lean.

When a machine is banked into a bend, trail gives rise to two opposing effects: (1) directional stability, tending to make the machine run straight and (2) the self-steering effect (also dependent on rake and wheel diameter) mentioned in Chapter 2, tending to steer the machine into the bend.

To achieve neutral handling, these effects have to be properly balanced; and the problem just mentioned is thought to be caused by an unsuitable combination of the two effects for that particular machine at a critical rake or trail. At the greater bank angles the self-steering effect would have outweighed the straight-ahead tendency.

Conclusions

1) Rake From our experiments it seems there is nothing magical in the conventional rake angle of 27 to 28 degrees. Indeed, balance, stability and lightness of steering were all enhanced by steepening the angle. The greater improvement came from the first change (from standard to 15 degrees), the subsequent move to zero rake producing only minor effects.

The only drawback noticed—juddering under braking—is a consequence of the poor structural integrity of the telescopic fork as a type. It is not suggested that machines should be built with a steep steering axis, using a headstock mounted fork, whether telescopic or leading/trailing link type, because the consequent high, forward location of the headstock causes structural and styling problems. Much better to consider some form of hub-centre steering or other wishbone layout, such as that used in the trail experiments.

2) Trail Apart from the need to avoid the critical situation mentioned, there seemed no obvious optimum value. Results were satisfactory throughout the full test range, so making personal preference the decisive factor.

The scope of our experiments was limited by time and money. Nevertheless, the results indicate a need for more exhaustive and quantitative testing. We hope that one of the large manufacturers may appreciate the potential benefits and divert some of their resources to further investigation.

Suspension

The primary function of suspension is to insulate both the rider and the bulk of the machine from road shocks—the first for his comfort, the second for mechanical reliability and longevity. In doing this, it is equally important to keep the wheels in the closest possible contact with the ground for maximum control. These functions can be studied under two major headings: springing and damping, and suspension geometry.

Springs

These can take many forms and be made from many materials but the practical range is more limited. Coil springs in steel are much the most common. They may be evenly wound (constant pitch) to give a linear rate; or they may be wound with a varying pitch to give a progressive rate. In that case, as the closer-spaced coils become coil-bound the effective number of coils is reduced and so the rate rises.

Leaf springs and torsion bars have been tried occasionally but have been superseded for various reasons although, in some applications, torsion bars may save valuable space. It is worth noting, however, that a coil spring is a torsion bar wound into a helix, though that makes it subject to bending stresses and stress concentrations that are absent in a straight torsion bar.

Titanium is an attractive material for springs, since it is twice as flexible as steel and little more than half the weight, size for size. Its disadvantage is high cost, so that its use is confined to such exotic machines as works racers.

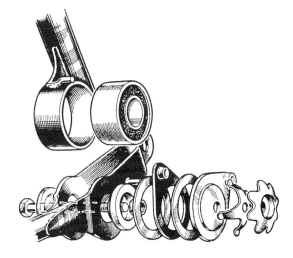

Rubber springing. Exploded drawing of Greeves front-fork pivot showing bonded-rubber bush. Although the bush was in torsion the rubber itself was stressed in shear

Rubber can be used in various ways. Greeves used it in the form of large bonded bushes that served as both the pivot bearings and the springing medium in their leading-link front forks. The rubber itself was stressed in shear, though the bush as a whole was loaded in torsion.

In Hagon telescopic forks, designed for grass-track racing, rubber bands provided the springing medium. The advantages here are light weight, low cost and ease of adjustment by adding or removing bands as required.

In this leading-link fork on a Motobecane scooter springing is provided by rubber bands in tension

Rubber has a natural progressive rate; it also provides some inherent damping, though this generates heat, especially in hard use. It is a versatile material and its characteristics can be tailored to suit different requirements by varying the composition and/or the mechanical design. On a mass-production basis, rubber springing lends itself to low unit costs. It merits consideration in any new design and may well be due for a revival.

To complete our survey of spring materials we must consider air or gas, which automatically provide a progressive rate. This can easily be demonstrated with a bicycle tyre pump. If you first extend the pump, then place a finger over the outlet and compress the pump, you will find the initial movement meets little resistance but the force required builds up rapidly with further movement.

Note also that if we operate the pump rapidly the air gets warmer. Since this is due to a fundamental law of nature we can't prevent it; all we can do in a pneumatic suspension system is to provide sufficient cooling to minimize the temperature rise, otherwise it will raise the effective spring rate (another natural law) unless we devise some form of compensation.

The extent of the progression in rate in a pneumatic strut is determined by its compression ratio (i.e. the ratio of the gas volumes at the two extremes of travel). The preload and initial rate depend on the static pressure and the area of the piston; in the case of air, this pressure can easily be adjusted with a tyre pump.

This was the case with the Dowty oil-damped pneumatic struts introduced by Velocette in the 1930s. Normal unladen pressure was 35 psi and the 3·25 in. stroke gave a high compression ratio, hence a steeply progressive rate. In more recent Japanese applications of pneumatic springing the air is mostly used to supplement steel coil springs.

Damping

This is necessary to prevent uncontrolled oscillations in the suspension. (Incidentally, it is incorrect to refer to dampers as shock absorbers, as we shall see.)

The simple harmonic motion we discussed in Chapter 2 applies equally to suspension. Imagine a large bump has fully compressed a suspension strut; at that instant energy is stored in the spring as potential energy.

As the spring returns to its static length it gives up this energy which, if there were no damping, would be transferred entirely to the mass of the bike (both sprung and unsprung) in the form of kinetic energy (energy of

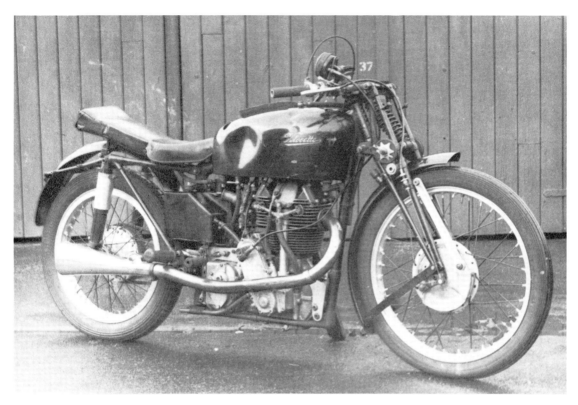

On this 1939 Mark 8 KTT Velocette the rear springing is by Dowty oleo-pneumatic struts inflated to a pressure of 35 psi. Stroke was $3\frac{1}{4}$ in.

motion). This would cause the suspension to extend well beyond its normal ride height and the whole process would be repeated.

If, however, we introduce the correct amount of damping, this overshooting will be prevented because, by the time the suspension returns to the static position all the energy stored in the spring will have been absorbed by the damper and dissipated as heat. This is why the dampers on hard-working motocross suspensions sometimes overheat. A damper is basically an *energy absorber* and should be matched to the rate of the springing medium.

Before the introduction of hydraulics, dampers were of the friction variety and their characteristics were precisely the opposite of what is required. The static frictional force (usually called stiction) was high; but once the damper moved, the friction dropped considerably. Since a damper can absorb energy only when it is moving, obtaining adequate damping involved excessively high stiction; hence a large force was needed to start the suspension moving and it was insensitive to small bumps.

By contrast, hydraulic dampers respond to a minimum of force, so providing sensitivity at low rates of motion, while high damping forces are available as the *speed* of the damper rises.

In a normal hydraulic strut there are two types of damping—viscous and hydrodynamic. Viscous damping arises from the shearing of the fluid and the force is directly proportional to the velocity of movement. Hydrodynamic damping is due to the mass transfer of fluid within the strut and the force is proportional to the *square* of the velocity. The proportions of the two types of damping in any strut depend on its internal design and fluid properties. There is also some mechanical friction (stiction) due to seals and bushes inside the unit.

Viscous damping is a mathematical nicety and can be matched precisely to a single-rate spring to give critical damping. In practice, however, a compromise must be struck, so that final matching of the damper to the spring is best determined by experiment.

Hydraulic dampers can be endowed with various characteristics by their internal design—e.g. one-way damping only; two-way damping with different rates for bump and recoil; dead spots in the movement and so on. Early on, one-way damping was common, on recoil only. The logic here was incomplete, suggesting that the wheel should be as free as possible to move upward in response to a bump, transmitting the minimum force to the sprung part of the machine for the rider's comfort; rebound was considered less important and so the damping was introduced there. As a result, for the same effect the strength of the damping had to be approximately double what it would have been if two-way damping had been used.

This sometimes gives rise to a serious problem when crossing a corrugated surface. Each bump compresses the suspension quickly (because of the absence of damping) while the subsequent recoil stroke is slowed by the heavy damping. This may prevent the wheel from returning quickly enough to maintain contact with the road, so that the suspension has not returned to its static position before the next bump. The rapid repetition of this action may soon ratchet the suspension into the fully compressed state, so giving the effect of a solid frame.

This simple form of damper shown in figure 3.1—in which a rod oscillates a piston in a cylinder of hydraulic fluid—suffers from a fundamental problem of displacement. In other words, as the damper is compressed the volume of the rod entering the cylinder reduces the space available for the fluid. Hence, a compressible medium (such as air) must be included to compensate. As the damper is shaken on bumps the air and oil mix to form a foam that drastically reduces the damping force.

One simple way to bypass this problem is to extend the rod through both ends of the cylinder so that there is no change in internal volume with piston movement. But this is not currently used for suspension damping and is confined to steering dampers.

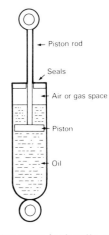

Fig. 3.1 **A simple two-way hydraulic suspension damper, in which some air or gas space (compressible) is necessary to allow for the piston rod entering and leaving the cylinder**

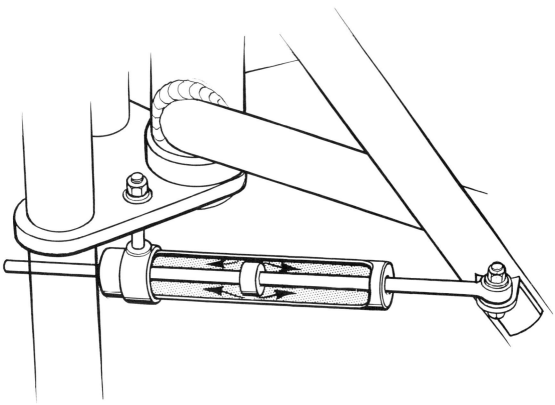

Fig. 3.2 **A hydraulic steering damper, with the piston rod extended to obviate the displacement problem. A typical installation**

← No need for gas space

← Rod extends through damper body

Many years ago, Girling adopted an ingenious solution to the displacement problem on their twin-tube strut. In this, the free air was replaced by a sealed nylon bag containing an inert gas Arcton (a trade name of ICI) and wrapped around the middle tube. Production, however, was slow and costly since the necessary absence of free air could only be guaranteed by a completely immersing each strut in hydraulic fluid before insertion of the bag and final assembly.

A later Girling solution was to revert to free gas but to substitute nitrogen at 100 psi for air.

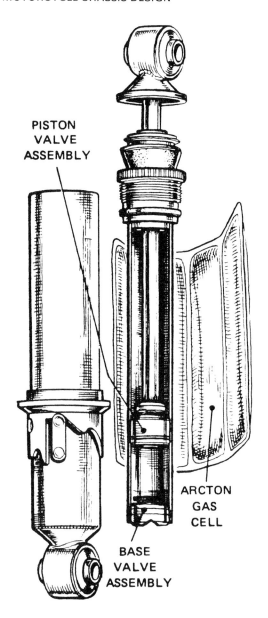

PISTON
VALVE
ASSEMBLY

ARCTON
GAS
CELL

BASE
VALVE
ASSEMBLY

Fig. 3.3 **To prevent aeration of the hydraulic fluid, the compressible element (Arcton gas) in this Girling strut was contained in a nylon cell**

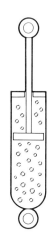

Fig. 3.4 **As a substitute for free air, nitrogen at 100 psi forms an emulsion with damping properties below those of the neat fluid, but which remain constant**

An emulsion was formed between the oil and gas, the resultant fluid giving constant damping.

Current design trends embrace variations on the nylon-bag principle and are loosely described as gas shocks. In these, the oil and pressurized nitrogen are usually separated by a floating piston, either in the hydraulic cylinder or in a separate compartment connected to it.

The gas pressure (up to 300 psi) keeps all the oil seals pressurized, so preventing leakage; it also helps stabilize the damping effect and prevents cavitation, however rapid the recoil stroke. On some dampers there is provision for adjusting the pressure as a means of fine tuning.

A recent refinement is the fitting of a simple external adjustment for the damping rate. This bypasses the need to remove the strut from the machine and the spring from the strut, as was the case with the earlier adjustment offered by Koni.

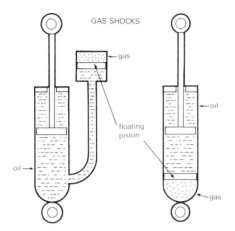

GAS SHOCKS

gas

floating
piston

oil

oil

gas

Fig. 3.5 In this later system the hydraulic fluid and pressurized gas (nitrogen at up to 300 psi) are separated by a floating piston, which moves to accommodate the displacement of the piston rod. In some cases the pressure chamber is separate from the strut and connected by a hose

Spring rate and wheel travel

We have to consider these together for the simple reason that the more wheel travel we have, the softer the springing we can use.

Over the years, wheel travel has increased steadily to the point where motocross machines have 10 to 12 in. This has enabled riders to negotiate rough terrain much faster for two reasons: first, the greater displacement can absorb larger bumps while softer springing imparts less movement to the sprung part of the machine, so enhancing control and comfort and thus reducing rider fatigue; second, the wheels are kept in closer contact with the ground, enabling more power to be transmitted (at the rear) and giving better steering (at the front).

However, extreme wheel travel may entail both mechanical and functional problems: in the first case, too much variation in chain tension or drive-shaft angularity; in the second, instability as a result of gross variations in

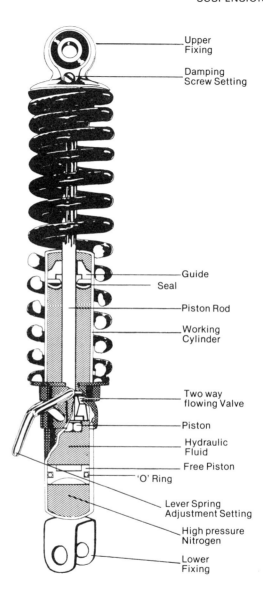

Upper Fixing
Damping Screw Setting
Guide
Seal
Piston Rod
Working Cylinder
Two way flowing Valve
Piston
Hydraulic Fluid
Free Piston
'O' Ring
Lever Spring Adjustment Setting
High pressure Nitrogen
Lower Fixing

In this De Carbon 'gas shock' the displacement of the piston rod is accommodated by a high-pressure nitrogen chamber separated from the hydraulic fluid by a free-floating piston. Spring preload and damping are both adjustable

steering geometry, and ride-height changes with different loads. Since variation of the load (rider's weight, passenger, luggage, fuel load) occurs mainly at the rear, this alters the rake and hence the trail.

Another consideration, often overlooked, is the effect of cornering forces, particularly on road-racing and sports machines, where bank-

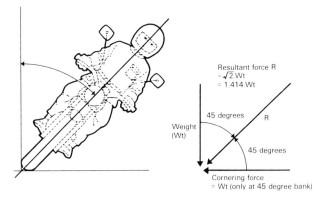

Resultant force R
$= \sqrt{2} .Wt$
$= 1.414 .Wt$

45 degrees

R

Weight
(Wt)

45 degrees

Cornering force
$= Wt$ (only at 45 degree bank)

Fig. 3.6 **At a cornering angle of 45 degrees (i.e. 1g lateral acceleration) the load on the suspension is 41 per cent higher than when the machine is upright**

ing angles may exceed 45 degrees. This represents a 40 per cent increase in the static suspension loading. (In ice racing, where machines are banked through 60–70 degrees, the static loadings could be increased two- or threefold, which may explain why rear springing has not been adopted in that sport.)

Imagine negotiating a series of opposite-hand bends at high cornering speed. As we bank into the first bend the suspension compresses under the increased loading; but when we straighten up and heel the other way it first extends as the cornering load is removed, then compresses again.

This not only reduces the suspension movement available in the bends for absorbing

bumps (when most needed) but also subjects the rider to an up-and-down movement that does nothing to improve his control. If the machine has telescopic forks there is a change in wheelbase too.

With single-rate springs, the rate needs to be soft enough to absorb the bumps but hard enough to prevent bottoming. In an attempt to improve the situation, some designers use progressive-rate springing, by means of either variable-pitch coil springs or a linkage that progressively increases the ratio of strut movement to wheel movement, so stiffening the effective rate.

This works well in motocross (where the main hazards are rough terrain and jumps) but may be a mixed blessing for road racing because the low initial rate leads to a disproportionate amount of the available wheel travel being taken up by cornering loads, leaving less travel *and* a higher spring rate to cope with any bumps.

The same considerations apply to a tourer, where the appreciable extra load of a passenger and luggage uses up the soft springing, leaving only the hard to handle bumps. In this case, though, the increased static load calls for a higher spring rate—but this can still be achieved only at the expense of movement and a change in attitude, unless the static ride height or spring preload are adjusted.

Ride height and preload

Most suspension struts incorporate an adjustment for the initial loading on the spring (preload). If there is some displacement in these struts under static load, then this adjustment will alter the ride height and can be used to compensate partially for different loads.

Some years ago, it was common to preload the suspension so that there was little or no displacement under static load and some force was needed to start the movement. The reason for this was that many riders found an improve-

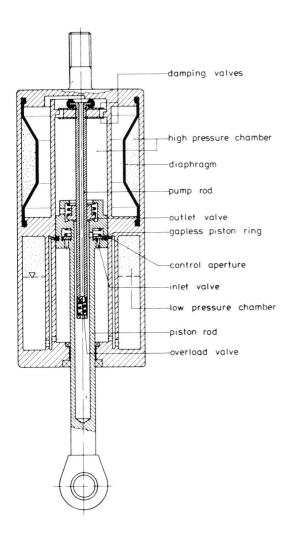

- damping valves
- high pressure chamber
- diaphragm
- pump rod
- outlet valve
- gapless piston ring
- control aperture
- inlet valve
- low pressure chamber
- piston rod
- overload valve

Fig. 3.7 **Section of Boge twin-chamber self-levelling suspension strut**

insufficient stiffness in the rear or front fork, in which case preload tended to stiffen these components.

It was these two shortcomings that gave rise to the popular misconception that hard springing is necessary for good handling.

Besides giving an uncomfortable ride, this approach restricts usuable suspension to *raised* bumps; with the springs preloaded, we are no better off in dealing with *hollows* than we would be with a rigid frame.

Mercifully, the current approach is to have stiff pivoted rear forks, efficient damping and softer springs adjusted to allow some extension of the struts on hollows. Ken Sprayson, designer of the famous Reynolds racing frames, used to specify one-third of the available wheel travel for extension and two-thirds for compression, which seems as good a starting point as any for racing.

An ingenious development for touring machines—designed to maintain ride height and suspension travel constant regardless of load—is the Boge Nivomat self-levelling strut, introduced on BMWs in the early 1970s. This is an extremely sophisticated hydropneumatic design using nitrogen as the springing medium and oil for damping. Unlike earlier self-levelling systems on cars, it requires no hydraulic pump and attendant pipework. Instead, an internal pump is energized by the suspension movement itself; this raises the internal pressure to bring the ride height up to a predetermined level. The pressure increase also causes an increase in the effective spring and damping rates.

This system can be used either on its own or in conjunction with auxiliary springs. The internal volumes and areas can be tailored to suit the characteristics desired in any particular machine.

ment in handling, albeit with a harder ride, when they adjusted the preload to the hardest setting.

This was a classic case of treating the symptoms rather than the illness, for the problem was twofold: 1) inefficient dampers, unable to control the available movement properly, hence it was helpful to restrict the movement (a self-defeating approach); 2)

It is to be hoped that this type of system will receive increasing consideration, as it solves many of the suspension problems arising from the relatively large variations in wheel loading common on a motorcycle, particularly at the rear.

Sprung and unsprung mass

When the front or rear suspension is partly or fully compressed, it exerts equal and opposite forces on both the sprung and unsprung parts of the machine—i.e. it tries simultaneously to raise the bulk of the machine and return the wheel whence it came. We want a system that returns the wheel quickly without unduly disturbing the sprung mass.

The acceleration of any object depends on its mass and the force acting on it. Since, in our case, the same force is acting on both the sprung and unsprung masses, their *relative* accelerations depend on the ratio of their masses, while their *absolute* accelerations are determined by the spring force and the absolute values of the two masses.

Hence, both for comfort (minimum disturbance of the sprung mass) and for roadholding (quick response of the unsprung mass) we need the highest possible ratio between the two.

Unfortunately, there is a limit to how light we can make the unsprung mass and so an increase in sprung mass enhances comfort and roadholding on rough roads. This increase requires a higher spring rate, which quickens the response of the unsprung mass.

This case and stability in side winds (discussed in the previous chapter) are the only two where mass is actually beneficial.

Front geometry

Except on scooters and some small utility machines, the telescopic front fork is almost universal. Its popularity cannot be justified on engineering grounds, however, because it is

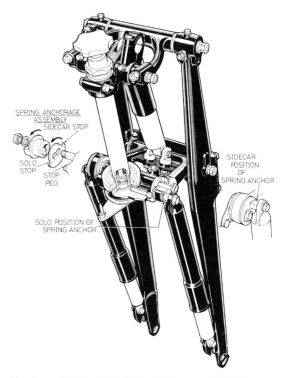

Most sophisticated girder fork was this Vincent Girdraulic, with forged light-alloy blades and one-piece upper and lower link assemblies. Trail was readily adjustable. Springs were in long telescopic tubes; hydraulic damper was separate. Lateral stiffness was enhanced by plate bridging front of blades (*MCW*)

technically bad from almost every point of view. Nor is low cost a valid explanation, as is sometimes claimed on the grounds that the main components are amenable to mass-production techniques. If cost was the decisive factor, it would be difficult to understand why the Japanese manufacturers fit link-type forks on their mass-marketed small commuter and shopping bikes, where a low selling price is even more important.

No, the chief reason for the telescopic's long reign is more likely to be a collective fear in the

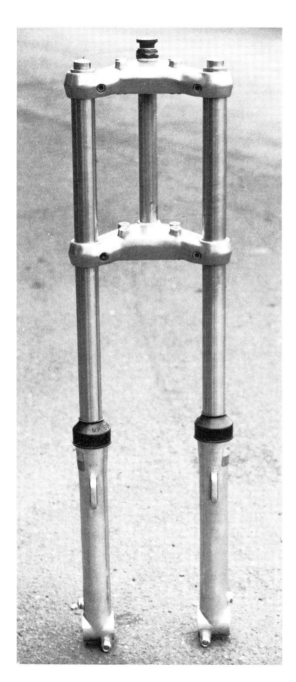

Left A typical telescopic fork by Metal Profiles. The sliders move outside the stanchions (*MCW*)

marketing departments of the major manufacturers that the fashion-conscious enthusiast is not yet ready to accept a change from such a visually appealing fork.

The modern telescopic fork comprises a pair of aluminium sliders fitted over chromium-plated steel stanchion tubes clamped in yokes at top and bottom of the steering column (see photograph). Although there was once a fashion for relatively large-diameter springs surrounding the stanchions, today's springs are usually of smaller diameter, longer and fitted inside the stanchions. A damping mechanism is incorporated in the sliders and the oil serves as a lubricant too.

Let us consider some of the defects. First, when the fork is fully extended there is minimum support for the sliders (because of the reduced overlap), so that the effect of the

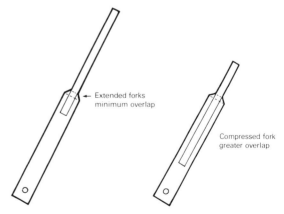

Extended forks
minimum overlap

Compressed fork
greater overlap

Fig. 3.8 When a telescopic fork is fully extended the sliders are poorly supported because of a reduced overlap on the stanchions. On full bump, overlap is at a maximum

Fig. 3.9 **Only if the front suspension alone were compressed on level ground, while the attitude of the rest of the machine was undisturbed, would the trail remain constant with a telescopic fork**

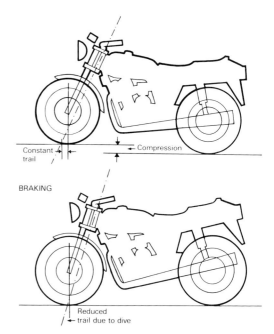

Braking reduces the trail as weight transfer compresses the front springs and extends those at the rear

working clearance is considerably magnified at the wheel spindle. Second, the sliders can move independently of one another except for the bracing effect of the wheel spindle at the bottom and perhaps a mudguard bracket at the top. Third, considering the loads and leverages imposed on them, the stanchions are quite small in diameter (typically 35–38 mm on large machines).

All these features add up to a fork that is relatively flexible in most directions; and, as we have mentioned earlier, lateral flexibility impairs stability.

Because of the rake angle, fore-and-aft loads are applied to the fork legs under static loading and this gives rise to stiction, which hardens the response to small bumps (see figure A2.3). Under braking, telescopic forks are well known for diving; and although this is commonly attributed solely to forward weight transfer, figure A2.3 shows that, for a normal rake angle, there is an additional factor at work—that is, a component of the braking force also acting to compress the legs. From the figure this component is $= F. \sin \Theta$, hence for a 27 degree rake this additional force is 45 per cent of the front wheel braking force.

Often it is claimed on behalf on teles that trail remains constant throughout the full range of wheel travel. This assumes, first, that constant trail is desirable (which is by no means certain) and, second, that there is no change in the attitude of the rest of the machine while the fork is compressed by level ground under the front wheel.

It is difficult to visualize any conditions under which this situation arises—or when both wheels move the same distance vertically. In practice, the fork is usually compressed

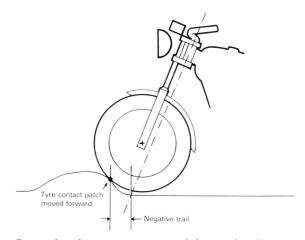

By moving the tyre contact patch forward, a sharp bump reduces trail (even from positive to negative)

either by braking nose-dive—in which case the rake angle is decreased, hence the trail shortened—or by hitting a bump, which initially moves the contact patch forward, also reducing the trail.

As the wheel moves over the bump, the trail first returns to its static value, then lengthens as the contact patch moves rearward, only to return to normal again as the wheel regains the level road. Considering these changes, we can hardly agree that a telescopic fork maintains constant trail.

Road bumps tend to impart a longitudinal force to the wheels as well as the vertical force expected. The rearward movement of the front

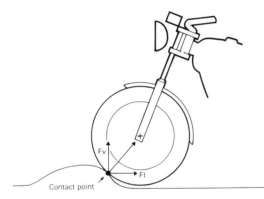

Fig. 3.10 The resultant wheel load from a bump acts radially inward from the contact point. Fv is the vertical component of the tyre force. Fl the longitudinal component

wheel under fork compression helps absorb this force, enhancing both comfort and control.

The Altec fork shows a valiant attempt by Alloy Technique, of Dartford, to eliminate some of the shortcomings inherent in more conventional telescopic forks.

The stanchions are extra long and extend through both ends of the offset sliders, thus supporting them fully throughout the full range of suspension travel and minimizing the

effects of the working clearance. The sliders are rigidly joined, not only by the wheel spindle but also by a strong bridge that virtually prevents the independent movement that is otherwise the chief cause of the tyre patch moving sideways.

A pair of interconnected hydraulic cylinders link the stanchions to the main frame in such a way as to permit free steering while resisting fore-and-aft flexure under heavy braking. (This also reduces the horizontal loads on the steering head, so reducing flexure in the frame and possibly prolonging its life.)

A single suspension strut, anchored to the frame, is connected to the sliders through a linkage system at the lower end; in this way the strut is no part of the steering inertia.

An alternative to the telescopic fork is the leading-link variety, which has the distinction of having been fitted to what were probably the best-handling racing machines to date— the world champion Moto Guzzis of the mid-1950s. The Earles fork, used on some early MV Agusta racers and for many years by BMW, is a variation on the same theme but has the disadvantage of considerably higher steered inertia.

Below Early postwar MV Agusta four with Earles-type front fork. Besides the high steered inertia, the fork had its bracing tubes positioned so as to contribute little to stiffness

Several examples of both varieties are shown in the photographs. Generally, they comprise a tubular or pressed-steel structure connecting the steering column to the link pivots and incorporating anchorages for the suspension struts. The links may be independent or formed by a single U-shape loop around the back of the wheel.

If the links are separate, then their resistance to independent movement, as in the case of the telescopic fork, depends on the rigidity of their attachment to the wheel spindle. In the better designs, this is of larger-than-usual diameter (hollow for lightness) and secured by extra-wide clamps. However, a large-diameter spindle means large wheel bearings and it may be that the most weight-efficient solution is a loop behind the wheel and a smaller-diameter spindle.

The benefits of this type of fork depend greatly on the quality of detail design. In most respects, a well thought-out leading-link fork is superior to the best of telescopics. Greater rigidity is possible, with benefits in stability and precise control. The lack of stiction considerably enhances sensitivity to small undulations and any degree of anti-dive under heavy braking can easily be designed in. The precise

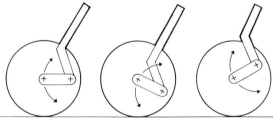

Fig. 3.11 **With leading-link forks, the relative heights of wheel spindle and link pivots determine the path of wheel travel**

path of the wheel travel depends on the relative heights of the link pivots and wheel spindle (see figure 3.11).

Trailing-link forks differ from the leading-link variety only insofar as the link pivots are ahead of the wheel spindle, not behind. Their

Left **A Foale leading-link fork with enclosed suspension units. For styling reasons the links were not joined; they were cast in light alloy with wide spindle housings for stiffness**

disadvantage is higher steering inertia, since the bulk of the mass is relatively far from the steering axis—an effect that is only partly off-set by the smaller amount of material needed to reach the pivots.

All the forks we have discussed—telescopic, leading-link and trailing-link—share a major disadvantage, i.e. they all need a high steering head mounted on the frame. This gives rise to

A leading-link fork which helped to create an enviable reputation for handling, as fitted to the 1954 NSU 125 cc Rennfox world championship-winner. The stanchions were fabricated from steel pressings and the links had wide spindle clamps (*MCW*)

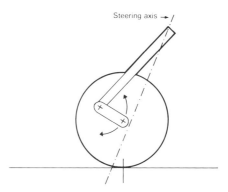

Fig. 3.12 In a trailing-link layout, steering inertia is increased by the relatively long distance of the mass of the fork from the steering axis

long forks which, through their leverage, can feed extremely high loads into the frame, so requiring it to be relatively heavy (see figure 2.21). Long forks are also a source of flexure, which deflects the contact patch to either side of the steering axis, to the detriment of stability and control.

Various attempts—some good, some bad—have been made to overcome these deficiencies. Several of them share common features grouped under the heading of hub-centre steering. Others defy such classification and must be considered on their individual merits. Let us now look at some of them.

Ner-a-Car

This early example of hub-centre steering had what amounted to a pivoted front fork (actually a pivoted U-shape axle, closed end forward) supporting an inclined kingpin in the centre of the wheel. The wheel swivelled on the kingpin and the swivel arm on the hub was

Left A very early example of hub-centre steering, on a two-stroke Ner-a-Car. Kingpin rake was **14 degrees** but varied with suspension travel, albeit short (*MCW*)

connected by a link to a 20-per-cent-longer arm at the bottom of the vertical steering column, so gearing-up the steering.

In its day the Ner-a-Car was renowned for its outstanding stability, though credit for that must be shared also by its long wheelbase (59 in. unsprung, 68·5 in. with pivoted-fork rear springing) and its exceptionally low centre of gravity. A drawback in the design was that the inclination of the kingpin (i.e. the rake angle) and hence the trail varied considerably with suspension movement, showing that constant trail is not all-important. With today's long, soft suspension that would probably be unacceptable but suspension travel on the Ner-a-Car was very short.

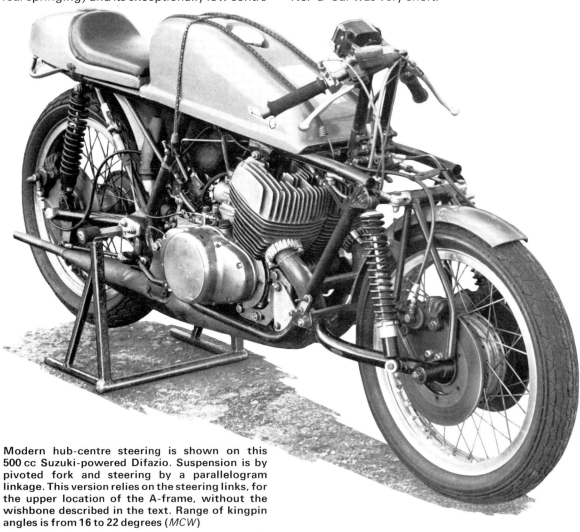

Modern hub-centre steering is shown on this 500 cc Suzuki-powered Difazio. Suspension is by pivoted fork and steering by a parallelogram linkage. This version relies on the steering links, for the upper location of the A-frame, without the wishbone described in the text. Range of kingpin angles is from 16 to 22 degrees (*MCW*)

There was no front brake on the Ner-a-Car but, if there had been, the effect on the suspension would have been the same as with more-conventional leading-link front forks: that is, if the shoe plate (drum brake) or caliper (disc brake) was anchored directly to the links, brake torque would have extended the suspension fully despite forward weight transfer, so causing juddering; but if the brake was anchored by a pivoted linkage to a sprung part of the chassis, then brake torque could have been isolated or any chosen degree of antidive built in.

Difazio

Since 1968 a hub-centre conversion designed by West Country engineer Jack Difazio has been available to riders of standard roadsters seeking to get away from the shortcomings of the conventional telescopic fork.

The axle, with a kingpin in the middle (raked between 16 and 22 degrees), is carried in a pivoted fork; to allow for suspension movement with minimal change in rake angle, the horizontal arm of the kingpin is bushed to rock on the axle. Swivelling on ball bearings at the top and bottom of the kingpin is a steerable but non-rotating drum supporting the large wheel bearings (130 mm outside diameter, 85 mm inside diameter); the drum is slotted for steering.

Bolted to the outer flanks of the drum are two upright A-frames (one each side) whose apexes are united by a bridge-piece above the wheel. On the 'Nessie' variant the middle of the bridge-piece is attached by a ball-joint to the forward apex of a substantially horizontal wishbone which, like the fork below it, is pivoted on the main frame. Steering is effected

Right **Difazio-type front end as used on Mead and Tomkinson 'Nessie'. Note wishbone location of upper end of the side A-frame** (*Mechanics*)

through a parallelogram linkage connecting the A-frames to the steering column.

Rake and trail are determined by the length and location of the wishbone. Since there is no offset, trail is related directly to the rake angle and wheel diameter. The brake calipers are mounted on the A-frames so that torque reaction is taken by the wishbone.

This design has the following advantages: 1) the degree of antidive is governed by the geometry of the fork, A-frames and wishbone (in-side view) and can be varied as required; 2) wheel movement is substantially upright, so maintaining an almost constant wheelbase; it is easy to arrange for the rake angle to remain constant too; 3) the trail can

be readily altered (with no change in wheelbase) by changing the lengths of the wishbone and steering links, so catering for different machines, road conditions and personal preferences.

A snag with any suspension system (front or rear) giving approximately vertical wheel movement is that this does not absorb the rearward component of the force due to road bumps, as does a telescopic fork. But this is probably a small price to pay for the benefits of hub-centre steering

Certainly, overall rigidity is much superior to that of a fork mounted on a conventional steering head. Also, since the loads fed into the main frame are much smaller, it can be made lighter. On the debit side, hub-centre layouts of the types described can be criticized on the grounds of appearance, weight and limited steering lock.

Other systems

Two French designs that may be loosely classified as hub-centre steering are those of André de Cortanze and Didier Jillet. Sponsored by Elf, Cortanze's design was first raced in Formula 750 with a Yamaha engine, then in endurance events with Honda power.

The general layout of this type of design is as follows: A car-type upright (a light-alloy casting or forging) supports either a fixed stub axle or a live rotating axle for one-sided wheel mounting; the brake calipers are bolted to the upright. Two one-sided longitudinal arms connect the upright to the main chassis via spherical ball joints. For unbiased steering (effected by a single drag link) the centre of

Early Yamaha-powered Elf. The front wheel is carried on a light-alloy upright connected to the main frame by upper and lower longitudinal pivoted arms. Steering is by a drag link. The offset between the wheel spindle and steering axis can be clearly seen (Dehesdin)

This Honda-powered Elf had different steering geometry from the Yamaha-engined model. At about 5 in. the trail was above average and resulted in heavier steering and a high degree of stability (**Edge**/*Bike*)

these joints must obviously lie in the centre plane of the wheel. A wide variety of steering geometries is possible.

Scrutiny of the photos indicates that the first Elf (with the TZ750 engine) had a fairly conventional steering geometry, there being some positive offset between the steering axis and wheel spindle combined with a normal rake angle, to achieve the desired trail.

However, the later Honda-engined design appears to use little offset, with a rake angle of approximately 23 degrees (although this might be varied for different circuits); this

gives a trail of about 5 in. This is higher than normal and probably accounts for the reported heavier than usual steering with good stability. The Jillet design has no offset and a steepish rake angle. Although the suspension strut could, in theory, be attached to either arm, its actual attachment to the lower one helps keep the weight low down; it is, of course, not steered.

This steering design is hard to fault and can have the following points in its favour: (a) low unsprung mass; (b) low overall mass; (c) high rigidity; (d) low steered inertia; (e) easy wheel changing; (f) reasonably neat appearance; (g) any constructional inaccuracies can be simply adjusted out by varying the lateral location of the pivoted arms or the pivot points of the uprights; hence wider tolerances are permissible in manufacture; (h) rake angle, hence trail, can be easily changed with minimal effect on wheelbase.

A valid criticism is that steering in one direction is restricted by the pivoted arms (albeit bowed for tyre clearance) and the steering drag link.

An interesting combination of telescopic fork and hub-centre steering was proposed and patented by Tom Killeen, who called it head-to-hub steering.

'The telescopic fork,' according to Killeen's description, 'is mounted on a kingpin or spherical bearing in the hub centre and on a self-aligning or spherical bearing at the steering head on the main frame. The arms of the pivoted fork are bowed outward for tyre clearance on full lock and the pivot on the main frame is so positioned as to minimize the rocking of the telescopic fork at the steering head on bump and rebound.'

Except in appearance, it is difficult to see what advantage this system offers compared with, say, the Difazio—but it is superior to the conventional telescopic fork in several ways. In all conditions other than braking, the sliders are relieved of fore-and-aft loading, hence stiction is minimal. Even during braking, these

A Ducati 900 version of the Foale system described on page 93. Rose joints locate the inner end of the wishbone. A simple drag link connects the handlebar to the fork for steering

longitudinal loads are only those resisting the caliper forces; the main loads are taken by the pivoted fork. The fork also resists differential movement of the telescopic legs and much enhances lateral stiffness.

It is sometimes claimed for hub-centre steering that the elimination of the conventional steering head reduces frontal area. In the case of a roadster, this is not so because of the upright riding position. But it is clearly true for a low record breaker such as the fully streamlined, 2 ft-diameter Harley-Davidson

Above **Hub-centre steering on Cal Rayborn's 1970 world-record Harley-Davidson projectile**

Below **By eliminating the conventional high steering head, Rayborn's hub-centre steering layout enabled the streamline shell to be kept down to 2 ft diameter (Weed)**

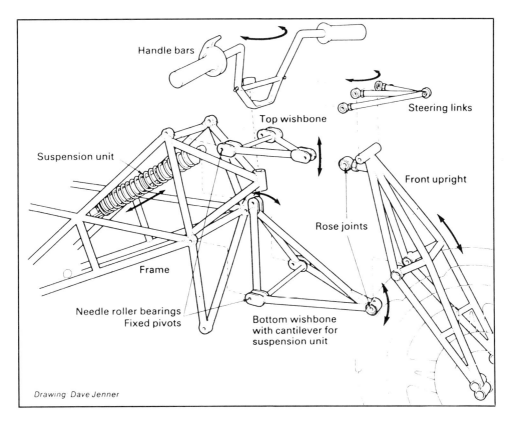

Handle bars

Top wishbone

Steering links

Suspension unit

Front upright

Rose joints

Frame

Needle roller bearings
Fixed pivots

Bottom wishbone
with cantilever for
suspension unit

Drawing Dave Jenner

Original Hossack tubular construction, showing front fork, steering linkage, wishbones and cantilever actuation of suspension strut. The steering linkage allows for suspension movement without introducing bump steer

projectile in which Cal Rayborn raised the world record beyond 265 mph in 1970.

At a casual glance, the Norman Hossack design (1980) shown in the sketch could be mistaken for an obsolete girder-fork layout. But any similarity is strictly superficial and the design is quite sophisticated and well thought-out.

Whereas, with a girder fork, the links are short and steered along with the spring, the only steered part of Hossack's suspension (apart from the wheel and brake) is the actual fork, which he calls the upright.

There is no steering head. Instead, the fork pivots on spherical bearings at the front apexes of upper and lower forward-facing wishbones. The lower wishbone is triangulated upward from its pivot to actuate the suspension strut. Rake and trail are easily altered by screwing the top spherical joint into or out of its lug at the top of the fork. A system of links designed to eliminate bump-steer connects the fork to the handlebar. The fork itself is rigid and well triangulated, so providing strong resistance to lateral deflection of the tyre contact patch from the steering axis.

As in the Difazio and Elf hub-centre layouts, the precise geometry of wheel movement is governed by the lengths and angular dispositions of the upper and lower wishbones, so allowing the designer much more flexibility than with a telescopic fork or one with leading or trailing links.

Indeed, the Hossack design may almost be regarded as an Elf with the pivot axes moved from within the wheel to above it. In this light, the Hossack has both advantages and disadvantages compared with the Elf. Among its advantages we can include: (a) the wishbones can be triangulated, so combining strength with lightness; (b) the layout of the wishbones is not dictated by tyre clearance, hence they

A later Hossack chassis, in which round tubes are replaced by fabricated box sections and the suspension strut is repositioned. Because the handlebar is close to the top of the fork the steering linkage is more complex than Tony Foale's single drag link (page 89)

do not restrict steering lock (c) a conventional wheel can be fitted, with twin brakes if necessary, so reducing costs for low production models and one-off specials; (d) appearance is more conventional, especially when a fairing is fitted, and air drag is reduced. On the debit side: (a) the higher location of the wishbones, fork, suspension strut and steering links raises the centre of gravity; (b) the lateral and longitudinal loads on the pivot bearings are increased by the extra leverage; (c) adjustment of rake and trail effects a greater change in wheelbase; (d) the main frame needs to be more complex, and possibly heavier, because of the longer load paths.

Though different in appearance, the Hossack system is similar in principle to a Foale design in which the upright, rather than being a triangulated fork, was fabricated in 2 × 1 in. box-section tubing, mainly for appearance and ease of construction. For a given standard of rigidity and strength, the triangulated construction can be lighter. Another, more functional difference, is in the steering geometry. To achieve the required trail, the Hossack design has a fairly normal rake angle and offset, whereas the Foale layout had no offset, hence a steeper rake of 10 to 15 degrees. See page 89.

Another Foale design combines the above system and Elf-type steering. In this, one of the steering pivots is within an inch of the hub centre while the other is just above the tyre.

One advantage that all the systems employing linkage enjoy over the telescopic fork (except for the Altec) is the opportunity to provide some form of geometric progression for the effective spring rate.

Rear geometry

In one form or another, the trailing pivoted arm (à la BMW 80 series and some scooters) or pivoted fork has long been established for rear suspension; and, although it is far from perfect, it is difficult to think of a better alternative.

With the notable exceptions of the Vincent and some early Moto Guzzi racers, it was long customary for the fork to comprise simply a cross-tube to house the pivot bearings, and a pair of side tubes to support the wheel and suspension struts. Clearly this elementary layout lacks torsional stiffness and preferably needs two suspension struts.

The most weight-efficient way to eliminate these defects is to triangulate the fork and con-

This Foale design uses a single suspension arm at the bottom and a wishbone at the top to support a box section upright. Hence, one steering pivot is just above the wheel, the other close to the hub centre. Steering lock to the left is 30 degrees, to the right 45 degrees. Suspension loads are fed up through the upright and top ball joint to the single damper and spring

nect the apex to the springing medium, as patented by Vincent-HRD in 1928. Nearly half a century later, Yamaha 'reinvented' the idea (initially for motocross), which acquired a semblance of novelty through the terms monoshock and cantilever.

Possibly to avoid a charge of copying, as well as to keep the weight low down, Suzuki triangulated the fork on the RG500 road racer below pivot level (as did Moto Guzzi in the mid-1930s), though retaining two shock absorbers. These were steeply inclined forward and fitted with progressive-rate springs.

Original (mid-1930s) Moto Guzzi pivoted rear springing. The fork is triangulated below pivot level and actuates long springs in horizontal boxes; upper links connect to friction-type dampers

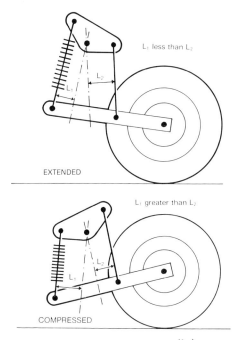

Fig. 3.13 Because the rocker ratio $\left(\dfrac{L_1}{L_2}\right)$ varies with wheel movement, this rear-suspension system gives progressive-rate springing and damping. The effective spring rate varies as the square of the ratio, i.e. $\left(\dfrac{L_1}{L_2}\right)^2$

However, any rising-rate effect would have been partly (if not wholly) offset by the geometric falling-rate effect due to the angled struts.

More recently there has been a trend towards rocker-type rear suspension, with a whole new jargon such as Kawasaki's Uni-track, Honda's Pro-link and Suzuki's Full Floater. The aim of all these designs is to obtain progressive-rate springing and damping by geometric means. If progression is desirable, then this is a good way to achieve it because the springing and damping rates vary together.

To achieve a progressive effect, we need to turn a link or lever through a large angle; for a given linear movement, this necessitates a short lever. All the rocker systems have this in common. Provided they give similar changes in effective spring rate (measured at the wheel spindle), and due regard is paid to stiffness and weight, then none of the layouts has any particular advantage over the others, despite the makers' claims. Thus the choice of design is best determined on structural and space considerations.

In the 1970s, the French Godier-Genoud endurance-racing Kawasaki was an early example of a link system, with the pivoted fork triangulated downwards and connected to the suspension strut via a bell-crank.

Bimota, in Italy, have tried various link systems, usually with a fork fabricated from box-section tubing and pivoted on the same axis as the final-drive sprocket to obviate changes in chain tension. Although this is a laudable objective, it involves an extra-long

Left This Godier-Genoud rear fork for a Kawasaki endurance racer is triangulated below pivot level. The link to the suspension rocker arm was connected to the lug at the front edge of the plate shown (*MCN*)

Below Bimota two-part chassis with the box-section rear fork pivoted on the chain-sprocket axis. The extra width and weight of the fork are a high price to pay to eliminate relatively small variations in chain tension. Later versions have reverted back to placing the pivot behind the gear-box (*MCW*)

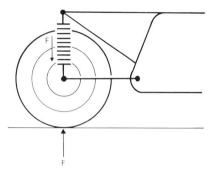

Fig. 3.14 **With the suspension struts mounted as shown there is virtually no load on the fork pivot**

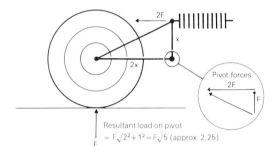

With a 2:1 leverage ratio, this triangulated fork produces a pivot load $2\frac{1}{4}$ times that on the wheel, in the direction shown

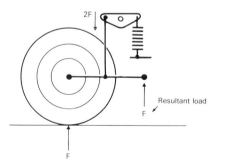

Also with a 2:1 leverage, this simple rocker layout produces a vertical pivot load equal to that on the wheel.
An infinite variety of systems is possible

fork, of considerable width at the front end. Unless an engine is specifically designed for this system, the extra width across the gearbox may be unacceptable; but the Bimota has rear-set footrests mounted where the width is less. It is interesting that the latest Bimota designs have dropped this system in favour of putting the pivot behind the gearbox.

Another method of maintaining more constant sprocket centres throughout suspension travel is to duplicate the fork to form a pivoted parallelogram on each side of the wheel, which is mounted between plates joining the rear ends of the upper and lower forks. The advantage here is that positioning the fork pivots further forward may not involve any increase in width. Against that, however, we must set extra weight, bulk and complexity.

A final consideration for a designer of pivoted rear suspension is pivot-bearing loads. With a simple fork, controlled by a pair of struts mounted upright at the rear, suspension forces place very little load on the pivot. However, the Vincent layout and all the rocker-arm systems considerably increase these loads. Figure 3.14 shows what happens in some selected designs.

In most cases, this may not constitute a serious problem; but bearing life may be shortened and the design of the pivot mounting on the main frame must take these increased loads into account.

Braking and driving effects

Depending on design, braking and driving forces and their internal reactions may cause the suspension to either extend or compress. For example, if we accelerate hard from rest on a shaft-driven machine, the rear end can be felt to rise as the driving pinion tries to climb up the crown wheel.

To put it another way, the crown-wheel housing is subject to an equal and opposite torque to that of the crown wheel itself. This

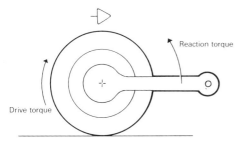

Fig. 3.15 **With conventional shaft transmission, the driving torque produces an opposite torque on the crown-wheel housing that tends to rotate the fork (or arm) backward, so lifting its pivot and rear of machine**

torque acts on the pivoted arm or fork, which therefore tries to rotate backward, so lifting the main-frame via the pivot and extending the suspension.

The effect can be reduced by mounting the housing in such a way that it is free to rotate on the axle, while taking the torque reaction forward to the main frame via a pivoted link approximately parallel to the suspension arm or fork leg. In this way the fork or arm is isolated from the torque of the housing; but since the angular motion of the housing is no longer the same as that of the fork or arm, it becomes

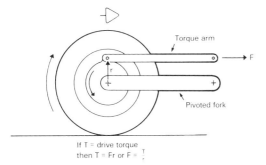

If T = drive torque
then T = Fr or F = $\frac{T}{r}$

Fig. 3.16 **If the crown-wheel housing is free to rotate on the wheel spindle and linked to the main frame by a pivoted torque arm, this prevents it from rotating the suspension fork (or arm) backward**

necessary to have a universal joint at the rear end of the drive shaft as well as the front, and to make provision for changes in effective shaft length.

Even so, this linkage arrangement cannot completely eliminate the effects of driving torque on suspension throughout the full range of travel. Figure 3.17 shows the effects of a parallelogram linkage at full bump and full recoil.

A certain amount of extending moment may be useful in counteracting suspension compression (squat) due to rearward weight transfer. But the system has been used on only a small number of specials, the major manufac-

Fig. 3.17 **Let F = drive force at tyre**
r = radius from wheel axis to torque arm
R = wheel radius
and let r = $\frac{R}{3}$ (typical value)

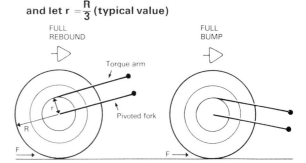

Then the balance of forces on wheel and housing is as follows:

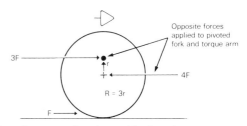

Now, because the pivoted fork and torque arm are angled, these forces create couples about the frame-mounted pivots

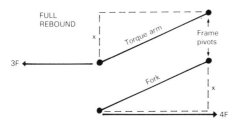

If suspension movement is symmetrical about the horizontal, then $x = \frac{1}{2}$ total movement

Extending moment = 4.F.x
Compressing moment = 3.F.x
Therefore, resultant extending moment = F.x (i.e. traction force times half full suspension movement)

Conversely, when the suspension is in the bump position, this moment becomes a compressing one. The only time the moment is zero is when the pivoted fork and torque arm are horizontal (i.e. mid-travel in this example). This effect can be altered by using non-parallel and/or unequal-length arms but we cannot eliminate the extending or compressing moments throughout the suspension range. If the total wheel movement is, say, 4 in. (then x = 2 in.) and the wheel's rolling radius is 12 in., then if the crown-wheel housing is fixed to the pivoted fork (as is usual) the extending moment is:

$$12.F \quad \pm \quad 2.F$$
Wheel torque   due to inclined pivoted fork

i.e. it varies between 10.F and 14.F.

From the example using a floating housing, the maximum moment is $\pm Fx$ or $\pm 2F$ (x = 2 in.), i.e. $\frac{1}{7}$ of the normal effect when extended or $\frac{1}{5}$ when compressed

Above This 1950 MV Agusta four-cylinder racer had shaft final drive and a rather elaborate way of isolating driving torque from rear suspension, using a parallel pair of pivoted forks and a floating crown-wheel housing

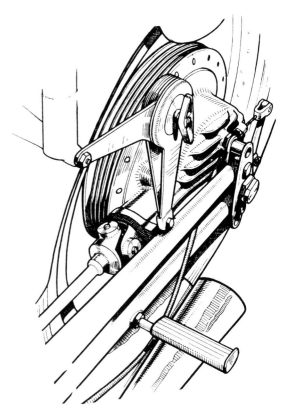

Right This drawing of the MV rear end shows the Hooke's joint at the rear end of the drive shaft; there was another at the front end and provision for slight variations in effective shaft length. Suspension damping was by friction

To prevent driving torque from extending the rear suspension excessively on this shaft-drive Guzzi the crown-wheel housing is free floating and connected to the main frame by a pivoted link. Note second Hooke's joint at rear of open drive shaft

turers presumably deciding that the extra complexity and cost would not be justified in terms of sales.

Nor is chain drive immune from influencing suspension. The magnitude of the effect in this case depends on suspension position, the difference in diameter of the front and rear sprockets and the relative lengths between centres of the sprockets and pivoted fork (or arm).

We can alter the effect by moving the suspension pivot up or down but this will increase the variation in chain tension with wheel deflection.

Braking reaction (rear)

The effect here is precisely opposite to what we described for a shaft-drive machine under

power—i.e. if the shoe plate or caliper is attached directly to the suspension arm, then the braking torque is applied to the main frame through this component, so tending to compress the suspension.

We can reduce this effect in the same way that we reduced the driving-torque effect— that is, by arranging for the caliper plate to float on the axle and taking the torque reaction forward to the main frame via a pivoted linkage of approximately parallelogram shape. Again, this reduces the effect to about 20 per cent of what it would otherwise be.

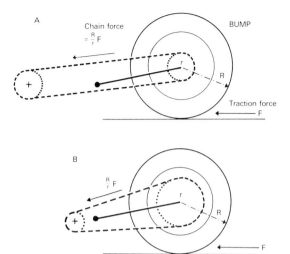

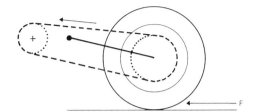

Sum forces on wheel spindle as before:

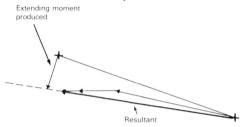

Fig. 3.18 Figures **A** and **B** show how the inclination of the chain relative to that of the pivoted fork varies with changes in sprocket size and fork-pivot position. Let us sum the forces acting on the wheel spindle:

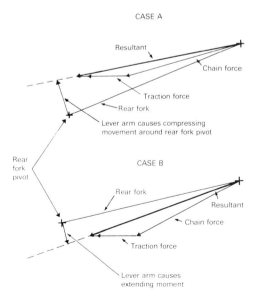

Under braking, some rear suspension compression may be useful in counteracting the change in the machine's attitude caused by forward weight transfer. But a serious problem may arise where there is no such parallelogram linkage (on most machines, in fact) and that is rear-wheel hop or judder under heavy braking.

The sequence of events bringing this about is as follows: Sudden application of the brake applies a sharp torque to the pivoted arm or fork, tending to compress the suspension; and, since the unsprung mass (of the wheel) is considerably less than the sprung mass, the wheel tends to leave the ground more quickly than the bulk of the machine tends to move downward.

As the wheel leaves the ground and the tyre therefore loses traction, the brake locks and so the compressing moment vanishes. When the wheel returns to the ground, still locked, it may skid; but the shock load on the tyre may cause sufficient sudden braking torque to set the whole process in motion again—and again.

Clearly, the best solution to this problem is the floating brake anchorage described; but

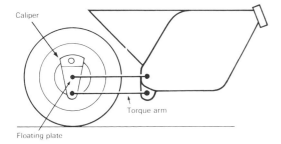

Caliper

Torque arm

Floating plate

Fig. 3.19 **To prevent brake torque from compressing the rear suspension, the caliper plate is free to rotate on the wheel spindle and linked to the main frame by a pivoted torque arm**

this is rare and it is more usual for manufacturers to tune the spring and damper rates to supress it. Unfortunately, this introduces an undesirable compromise into the selection of suspension characteristics, which are best determined by considerations of roadholding and comfort.

Because rear-wheel hop is most likely on very rough surfaces, it is on motocross and enduro bikes that a floating brake anchorage is most often found—Husqvarna have employed it for many years with others following suit later. In road racing, adoption of the scheme has been slower although, as with other sound features, a precedent was set long ago by Ing. Giulio Carcano on the grand-prix Moto Guzzis. Riders with experience of a floating brake anchorage testify to its worth despite the slight complication.

Before we leave the rear end let us consider the experiment conducted by Aermacchi Harley-Davidson on their racing two-stroke twins in the mid-1970s. They mounted the brake disc on the gearbox sprocket so that the braking force was taken through the chain. The benefits were: 1) less unsprung mass; 2) less torque on the pivoted fork; 3) a smaller and lighter disc, since the secondary gearing increased its rotational speed. Offsetting these

advantages, the disc was more shielded from the cooling breeze and chain slack could cause brake judder. If the bottom chain run is slack when the brake is first applied, little load is applied to the disc, which may momentarily lock or be slowed considerably below the corresponding road speed. Then, when the chain run is tightened by rear-wheel rotation, a shock load is applied to the wheel, so leading to an oscillation between disc and wheel, hence to judder. This may well be the reason the idea was abandoned.

Braking reaction (front)

We have already dealt with this in relation to the telescopic fork, where the main effect is compression of the springs by the resolved component of the braking force acting along the fork legs. Because of the severe nose-diving inherent in this type of fork, various attempts have been made to counteract the effect.

A recent system used by some Japanese manufacturers, on both roadgoing and racing

Unusual rear-brake arrangement on the Harley-Davidson two-stroke racing twins of the mid-1970s: the perforated disc was mounted on the gearbox sprocket rather than the wheel. The pros and cons are detailed in the text

machines, is to interconnect the hydraulic braking and damping systems in such a way that fork damping is increased on brake application. This slows the rate of nose-diving but is, at best, a poor solution to a problem of poor fork design.

A much better approach, albeit visually more complicated, is shown on page 104.

The caliper is mounted on a bracket, which is free to rotate on the wheel spindle, and the torque reaction is taken by a pivoted rod connected to the sprung part of the fork (usually below the lower yoke). On brake application,

Above This picture of Barry Sheene on a Suzuki RG500 grand-prix machine shows the nose-diving effect of weight transfer when braking hard with a telescopic front fork (*MCW*)

Right Suzuki's method of counteracting nose dive. A bleed from the brake hydraulic line operates a valve in the front fork to stiffen the damping considerably. Note also the floating brake disc operating surface with rigid carrier bolted to the hub (*MCW*)

Anti-dive system on the Kawasaki KR500 grand -prix machine. The free-floating caliper bracket is connected by a link rod to the lower fork yoke (*MCW*)

the force in this rod tends to lift the front of the machine against forward weight transfer. The magnitude of the lifting effect can be controlled by the distance from the wheel spindle to the attachment of the torque rod. This system has the added advantage of relieving the fork sliders of caliper forces, so reducing stiction and improving response to the road surface.

Because of its less tidy appearance, this device is unlikely to be used by the major manufacturers for their road machines; as far as is known, it has been confined to a small number of competition bikes. Technically, however, it is a superior solution to that previously described.

Link forks

On earlier BMWs, where the brake shoe plate is directly clamped to the Earles-type pivoted fork, the effect is the reverse of what happens at the rear—i.e. brake torque tends to extend the suspension in opposition to the effect of forward weight transfer. Even if full extension is not achieved, the suspension is effectively stiffened and so loses its sensitivity to road shocks. Another problem here is loss of feel— some degree of dive seems necessary to aid the rider's judgement.

On forks with shorter leading links, it is usual to incorporate a floating brake anchorage to prevent full fork extension, while any desired degree of antidive can be designed in by manipulating the link geometry.

Some scooter forks have short trailing links to which the shoe plate is directly anchored. In this case the torque reaction on the links reinforces the effect of forward weight transfer to give severe nose diving.

Hub-centre and similar

The double-link system used in most of these designs is geometrically similar to that of leading-link forks with floating brake anchorages. Again, antidive can be built in, though other considerations may influence the layout.

Structural considerations

Earlier chapters have detailed the effects of the various geometric parameters involved in chassis design and noted that these must be maintained by the structures concerned under all the load conditions to be expected in use. Moreover, the structures must be as light as possible consistent with an acceptable life span.

First we have to define stiffness and strength. Stiffness is concerned with the *temporary* deformation of a structure when loaded and unloaded and is measured in terms of the linear or angular flexure compared with the force or torque applied. Strength is a measure of the loading that can be applied before structural failure occurs. This failure may be either breakage of some part or *permanent* deformation which remains after the load is removed.

Fatigue

Failure rarely results from the static application of normal operating loads. Rather it is due to either excessive loading (such as a crash, which may result in breakage and/or permanent deformation) or fatigue, which ultimately leads to breakage.

Metal fatigue is a very important and complex subject, beyond the scope of this book. Suffice it to say that, if a motorcycle chassis is subjected only to normal operating loads, fatigue will be the most likely cause of failure. The essence of good design in this respect is to ensure that fatigue failure would occur only long after the expected life of the machine.

Fatigue results from continual stress reversal; an extreme example is the fracture of a piece of metal by bending it back and forth several times. In practice, the stress levels in a structure will be such that many millions of reversals are required to cause a breakage.

Fatigue characteristics vary from metal to metal. Some metals, such as steel for example, have a stress limit below which they will not fail no matter how many reversals they sustain. Some other metals, such as aluminium and its alloys, on the other hand, will eventually fail as a result of stress reversals, no matter how small the stress, although at low stress levels the number of reversals required to produce failure will be extremely large.

Consequently, great care is essential when contemplating a frame design in aluminium, since failure is almost inevitable if it is used long enough. Porsche, indeed, for their famous sports racing cars—with tubular space frames in aluminium and magnesium—pursued a policy of building new cars for every important event.

Stress reversals in a motorcycle chassis can be caused both by road irregularities and by engine vibration, which can give rise to very large forces.

Structural efficiency

If the components of a chassis are designed to be sufficiently rigid then, provided sound practice is applied to the details, strength will not usually to be a problem. Hence, a good guide to the efficiency of a structure is its

Triangulation in nature: a section through a bone in a bird's wing, giving a high level of stiffness to weight

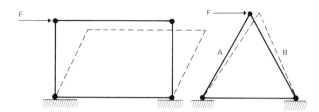

Fig. 4.1 Under force F, the four-sided structure easily distorts to a lozenge shape, relying on corner stiffness alone to prevent complete collapse. For the triangular structure to distort as shown would require side A to lengthen and side B to shorten

stiffness/weight ratio. In mass production, however, with common materials, cost is closely related to weight and so a major manufacturer might rather measure structural efficiency by the ratio of stiffness to cost.

There are two basic routes to structural efficiency (as defined by stiffness/weight). One is to use many small-diameter straight tubes in a triangulated frame; the other is to use few large-section tubes and rely on their inherent torsional and bending stiffness.

(A fine example of triangulation in nature, where efficiency is essential to survival, is the bone structure of some birds' wings.)

Triangulation

To visualize the effect of triangulation, we need only consider the two simple structures illustrated in figure 4.1.

If their bases are fixed while a force is applied as shown, then the four-sided frame may distort to a lozenge shape, with complete collapse prevented only by the tubes' resistance to bending at the corners. In contrast, the triangular frame can distort only by a change in length of any or all three sides. A few experiments with a piece of wire or welding rod will show how easy it is to bend but how difficult it is to change in length, thus proving how

much more rigid the triangular arrangement can be.

A practical structure such as a motorcycle frame may comprise several such triangles, and, if designed correctly, should be very efficient.

To test whether a design is fully triangulated or not, imagine that all the connections are by pin-joints—i.e. the joints have no resistance to bending. If the structure remains intact when loaded, then it is fully triangulated and may be regarded as a complete structure. If it collapses, however (as would our four-sided figure), then it is structurally unsound and is called a mechanism.

Of course, the four-sided frame may be stiffened tremendously by adding one or two diagonals (known as bracing struts), so converting it to two or four triangles.

If only one bracing strut is added, it must be of sufficient diameter to resist compression loads if the direction of the applied force can be reversed. But if two diagonals are added they can be in very thin material, even wire, because one or other will be subject only to tension and this one will complete the structure. The classic example of this technique is the wire bracing between the upper and lower wings of early biplanes.

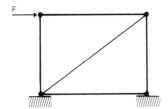 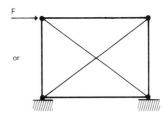

Fig. 4.2 In the pin-jointed structure on the left, the diagonal bracing strut provides effective triangulation. Two diagonal struts (right), however thin, stiffen the structure in both directions since one is always in tension

Referring back to our pin-jointed structure, it is obvious that bending moments cannot be fed into the individual members, which are thus subject only to tension or compression. if we know the magnitudes and directions of the applied loads, it is a straightforward matter (albeit involved) to calculate the cross-sectional area of the metal needed to withstand the individual member loads.

If the load in a member is only tension, then the cross-sectional shape and size are irrelevant. But under compression a member may tend to buckle, particularly if it is long (note the difference between pulling and pushing a piece of string!). In that case buckling is best prevented by using a round tube of large diameter and thin gauge.

Motorcycle frames comprising bolted triangular structures have been built and they resemble our theoretical pin-jointed example. Although this approach has some advantages, such as simplified repair, there are disadvantages, too. Movement may develop in the joints—as a result of wear, corrosion or even manufacturing clearances—and this will negate the stiffness benefits of triangulation. Also, the nuts and bolts at the joints may weigh more than welding.

Indeed, it is more usual to weld the joints, though this too can cause problems. In prac-

tice, it is rarely possible to ensure that all the loads in the individual tubes are either compressive or tensile. The very thickness of the tubes introduces physical offsets from the perfect triangular form, which in turn creates bending moments that are sometimes impossible to calculate.

To complicate the issue further, the welding process itself will leave indeterminate residual stresses in the chassis, while the joints themselves may cause stress concentrations (see later). All these considerations mean that higher safety factors may have to be applied to the calculated stresses if fatigue life is to be adequate.

In laying out a triangulated frame, simple considerations of space and shape may present difficulties, since some engines just don't lend themselves to that type of construction. Nevertheless there are one or two ways of getting out of trouble. For example, three additional solutions to the problem of stiffening a four-sided structure where there is no room for a direct bracing strut are shown in figure 4.3.

The second and third methods are examples of what might be called external triangulation—and the second is the basis of the Vincent-type of pivoted rear fork.

An example of the use of external triangulation is the Bimota KB2 frame, in which the steering head is supported by several tubes. As a complex chassis, involving many tubes and much welding, this would not be a viable proposition for a large manufacturer; but it is acceptable in the context of Bimota's specialist

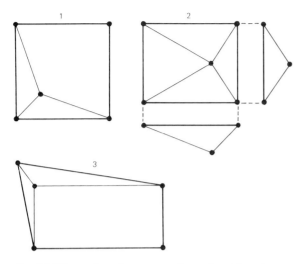

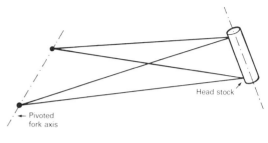

Fig. 4.4 **This simple frame, triangulated in both horizontal and vertical planes, is seldom practicable**

Fig. 4.3 Where there is no room for a direct bracing strut, a four-sided structure may be stiffened by any of these forms of triangulation

output. Indeed, such is its visual impact that it may have marketing as well as technical merits.

For a machine with a conventional steering head it is possible to design a simple triangulated structure connecting the steering head to the rear-fork pivot, as shown in figure 4.4.

However, this is seldom practicable because of the modifications necessary to accommodate the engine, seat and other com-

This sketch of the Bimota KB2 chassis shows the considerable use of external triangulation and the multi-tube support for the steering head. The large amount of welding involved restricts such a layout to small-scale production

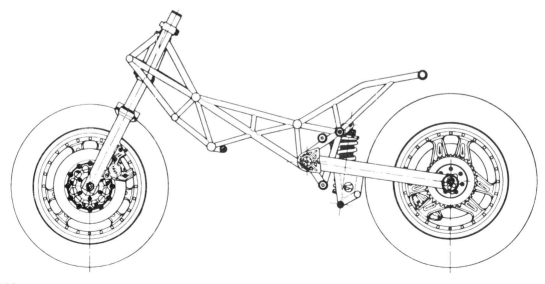

In this Foale frame for a 250 cc Rotax-powered racer a simple triangulated structure connects the steering head to the rear-fork pivot. Not many engines lend themselves to this arrangement

ponents. Nevertheless the Foale frame for a 250 cc Rotax-powered racer was built on these lines.

Beam frames

This heading covers several different types (e.g. tubular backbone, pressed-steel beam, monocoque) that use large-section members for their inherent rigidity under torsional and bending loads. Beams can also be combined with triangulation to produce a practical layout.

To illustrate the relationship between tube size and stiffness, let us consider two equal-length (L) segments of round tubing which are also equal in weight but of different diameters. Clearly, since the weights are the same, the larger tube has a proportionally thinner wall. If these tubes are subjected to a force (F) as shown, tending to bend them, the bending moment at the base of the segments is $F \times L$. Imagine the bases to be pivoted at point O and the bending moment resisted by a force at the centre (f_1 and f_2). To balance the applied moment, f_1 and f_2 are inversely proportional to the tube diameters—i.e. if the larger tube is twice the diameter of the other, then f_2 will have half the value of f_1. See figure 4.5.

Now let us assume that the deflection at the centre is proportional to this force—i.e. double for the smaller tube. But since this double deflection is at only half the radius, the net effect is that the *angular* deflection of the smaller tube is four times that of the larger tube.

This angular deflection translates to a lateral movement at the top of the member; thus, for the same loading, the deflection about O of the top segment of the double-diameter tube is only a quarter of the deflection of the other. The same reasoning applies to all the segments of equal-length members; therefore the resistance to an applied lateral load is four times as great when tube diameter is doubled.

Although the above is a simplified method of considering the problem, a mathematical analysis gives the same results: the lateral stiffness of equal-length, equal-weight, thin-wall tubes of the same material is proportional to the square of the diameter. Torsional stiffness follows the same rule. And, although we have used round tubes to illustrate the principle, it holds for all other cross-sectional shapes.

For readers with more mathematical knowledge, a tube's stiffness depends on what is usually called the moment of inertia (but should more correctly be called the second moment of area) of the cross-section about the bending axis (known as the neutral axis).

If, in the pursuit of structural efficiency, we were to follow this large-tube route to its logical conclusion, we should arrive at very large sections with walls little thicker than foil. Such ultra-thin tubes would buckle and collapse under load—and so, in practice, we have to compromise between wall thickness and tube diameter.

(Modern large aircraft are an example of the extreme use of the large-tube principle, with an outer shell of thin aluminium sheeting as the main structural member. But in their case buckling is prevented by bulkheads and many stiffening ribs supporting the shell.)

In a beam-type frame, *local* buckling is another pitfall that must be avoided by paying great attention to the detail design of the method of feeding in the loads. It is possible to achieve adequate overall rigidity, while still having local weakness that can lead to fatigue failure.

Before turning to the practical application of these principles, we must consider two more important structural concepts—the neutral axis and bending stress.

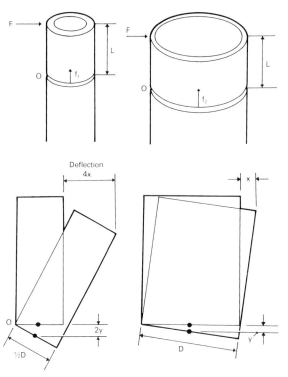

Fig. 4.5 **Relative stiffness of equal-weight tubes of different diameters**

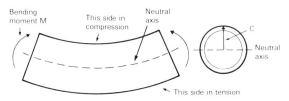

Fig. 4.6 **In a member subject to a bending load, the inner surface is compressed and the outer surface stretched. At the neutral axis (between the two) length is unaltered by bending**

Figure 4.6 shows a side view of a structural member subject to a bending load; this shortens the top surface (subject to compression) and stretches the lower surface (subject to tension). Somewhere between the two surfaces is a position of zero length change and this is known as the neutral axis. The bending produces no tension or compression stresses on this axis—unlike the outer surfaces, which are subject to stress that can be calculated by the following formula:

$$_{max}\sigma = \frac{M.c}{I}$$

where σ max is the maximum fibre stress
     M is the applied bending moment
     c is the distance from the neutral axis to the farthest fibre
     I is the second moment of area about the neutral axis

From this it follows that, to keep surface stresses to a minimum, we must keep c as small as possible consistent with a large value for I. This requirement highlights a danger inherent in the use of sheet-metal gussets to stiffen a tubular chassis, especially where the tubes are of large diameter.

Imagine that the frame section illustrated, with a welded-in gusset, is subject to a bending moment. If the gusset is relatively thin, then the

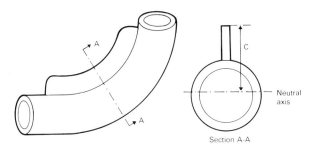

Fig. 4.7 **Thin ribs or gussets may increase stresses. Unless I (second moment of area) is increased in proportion to c, the maximum fibre stress in the rib will be higher than in a plain tube**

increase in I is small and the neutral axis is moved only slightly, hence c is relatively large; this increases the maximum stress, compared with that in a plain tube, and thereby increases the chance of failure. (NB: this higher stress is only on the outer edge of the rib; the tube stress is not raised.)

If the bending moment puts the gusset in compression, the failure will probably be buckling of the gusset; but if the gusset is in tension, then the failure will constitute tearing or cracking, which may spread into the tube itself and cause total failure.

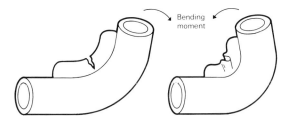

Fig. 4.8 **Two types of gusset failure: tearing under tensile stress (left) and buckling under compression**

Unless 'strengthening' gussets or ribs are sized correctly, therefore, there is a risk that they will actually weaken the structure. The same applies to ribs on castings.

Another danger in the use of ribs, brackets and other causes of sharp changes in section is stress concentration.

To explain this, figure 4.9 shows a solid bar subject to tensile stress, which may be represented by a series of equally spaced 'force lines'. When these force lines meet a sudden change in section, such as a notch, they bunch up in the vicinity to raise the *local* stress to two or three (or even more) times the average across the section. Any change in section can cause stress concentration and increase the failure risk, the degree of which depends on the sharpness of the change in section and the

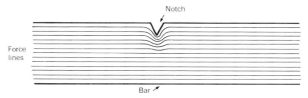

Fig. 4.9 The abrupt change in section at the notch causes the force lines to bunch up and increase local stress considerably, so shortening fatigue life

ductility of the material; but *any* sudden change increases the risk of failure and this is very important to the fatigue life of a chassis.

Typical areas of stress concentration are any attachment points such as the ends of gussets and brackets, the edges of welded joints and any places where a relatively flexible part of the structure is joined to a more rigid part. This explains the popular fallacy that a stiff frame is more inclined to break than a flimsy one. A gradual change in flexibility is required.

To avoid problems arising from stress concentration, the designer should as far as possible position brackets, gussets and so forth at areas of low basic stress. This can often be done by welding them to the neutral axis of the supporting tube.

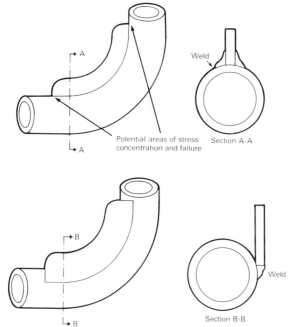

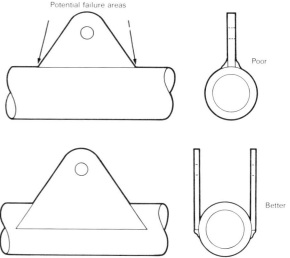

Fig. 4.10 Wrong (top) and right ways to weld a gusset on to a bent tube

The same principle applies to brackets welded to straight tubes

The sort of techniques illustrated can prolong fatigue life greatly.

The preceding description of various structural effects provides an essential basis for a sound understanding of practical chassis design. Of necessity, the descriptions are brief and simple. Any reader wanting to study the subject in greater depth is recommended to read any of the many books on Strength of Materials. Now let us consider the more practical aspects of various basic frame designs.

Triangulated frames

Although these can have extremely high structural efficiency, they have found few adherents among the major manufacturers. Probably this is because the shape and size of the most popular engine types require a wide and complicated (hence expensive) structure.

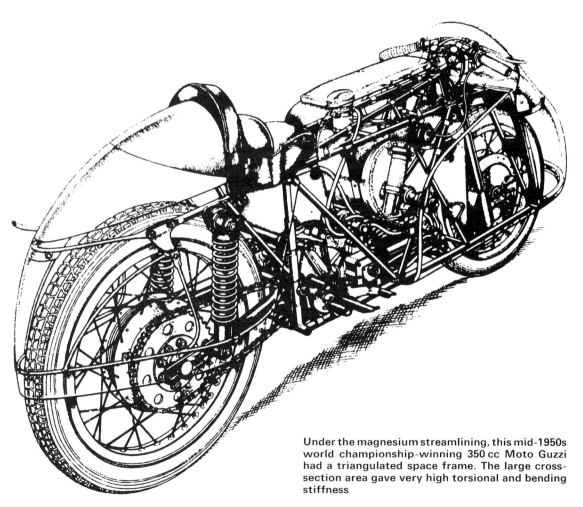

Under the magnesium streamlining, this mid-1950s world championship-winning 350 cc Moto Guzzi had a triangulated space frame. The large cross-section area gave very high torsional and bending stiffness

Left Tony Foale's own 350 racing Aermacchi of the early 1970s. Features a triangulated frame which also uses the engine as a structural member. The alloy plates at the rear of the engine support the rear fork pivot. The rear fork is triangulated with a single suspension unit above the engine

Bottom **Triangulation** *in extremis*: a Krauser chassis for a roadgoing BMW transverse flat twin. Its high price reflects the large amount of labour involved in welding the many joints

Frames of this sort have found most favour with the Italians, having been used on several Moto Guzzi works racers in the 1950s and subsequently on the 500 cc Linto twins. In both cases, the engines had horizontal cylinders and so presented no great accommodation difficulties. Other examples in the ranks of racing machines and low-production specialist roadsters include a few Norton racers in the early 1970s, the Krauser BMW roadster and Bimota KB2 chassis.

A problem to watch out for with long tubes of small diameter is engine-excited resonance—that is, severe vibration in the tubes caused by unbalanced engine inertia forces at a critical frequency. The solution is to raise the tube's natural frequency, either by shortening it or increasing its diameter. This phenomenon is not unique to triangulated frames; it can occur in any design using long, thin members.

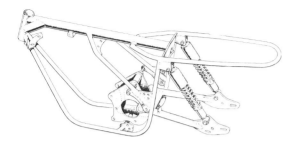

The Norton Commando frame used a tubular backbone and some triangulation but sacrificed some of its potential stiffness through the rubber mounting of assembly carrying the rear-fork pivot (*MCW*)

Inexpensive Simson 50 cc commuter bike features a small-diameter curved backbone frame to keep production cost down. The engine was cantilevered from rear crankcase rubber mountings

Tubular backbone

This type of frame too has failed to achieve the wide acceptance its structural efficiency merits (provided the backbone has a large enough diameter). Again there may be difficulty in accommodating bulky engines. Ideally, the tube should be straight and connect the steering head directly to the rear-suspension pivot; but in practice this is seldom possible. With flat or medium-size engines, however, it is usually feasible to bring the backbone to within a few inches of the pivot and bridge the gap with a welded-up box section. Another way to gain the necessary engine clearance is to bend the tube.

Even where these methods are unsuitable, an efficient frame can still be produced by using a high-level straight backbone and linking it to the engine and rear-suspension pivot by means of an arrangement of smaller-diameter straight tubes—provided these are properly triangulated (in both fore-and-aft and sideways planes).

The Norton Commando frame was of this type but sacrificed some of the potential stiffness because the engine-gearbox assembly on which the rear fork was pivoted was itself rubber mounted to the frame.

For simplicity and low manufacturing costs, many mopeds have a curved backbone of smallish diameter. A typical example is the MZ Simson 50, where the subframe supporting the seat is bolted on, so facilitating both factory assembly and accident repairs.

Some designers have combined a curved backbone with use of the engine as a structural member. An example was the 125 cc Honda twin that spearheaded the Japanese invasion of the TT in 1959—the small-diameter backbone being curved through nearly 90 degrees and the engine bolted in to stiffen the assembly.

Fabricated backbone

Quite a range of designs is possible here—the most popular being a T-shape structure comprising left and right steel pressings united by spot or electric-resistance welding, as in the Ariel Leader and Yamaha FS1E shown here. This construction makes for rigidity and low production cost, though the high initial tooling

115

Above **This stripped TT version of a 1960 Ariel Arrow (shown here by tuner Herman Meier) clearly shows the T-shape of the fabricated backbone frame, welded from left and right pressings** (*MCW*)

Left **The popular Yamaha FS1E is another example of a lightweight with a fabricated backbone for rigidity and low-cost mass production**

outlay rules it out for small production runs and specials. Also, the end product is heavier than an equally rigid tubular backbone because of the inevitable excess metal in areas of low stress.

Other notable variations on the fabrication theme include the 250 cc Ossa grand-prix single of the late 1960s and the later Kawasaki KR500 square four. The Ossa had a complete welded-aluminium chassis that doubled as a fuel tank, while the Kawasaki had a 32-litre fuel tank as a backbone, with the steering head housed in the front and two inner and two outer aluminium side plates attached to the rear to support the fork pivot and engine.

On Santiago Herrero's 1960s 250 cc grand-prix racer the welded aluminium chassis doubled as a fuel container (*MCW*)

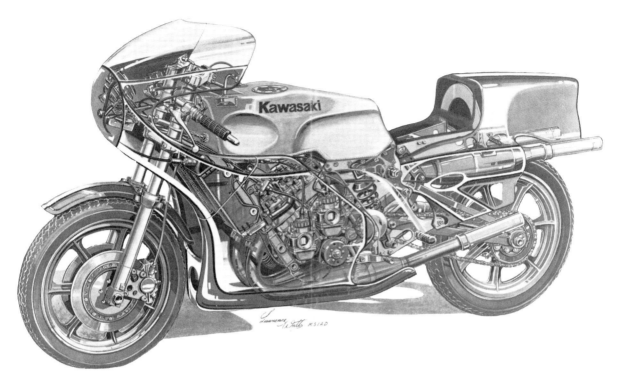

The fabricated light-alloy backbone on Kork Ballington's Kawasaki KR500 grand-prix machine comprised a 32-litre fuel tank with the steering head incorporated at the front. Aluminium side plates at the rear supported the engine and fork pivot (*MCW*)

Monocoque

This term, often misused when applied to motorcycles, was originally coined to describe aircraft that used a skin of sheet aluminium as both the structure and smooth outer shape; it was later applied to cars employing a similar technique.

A motorcycle, however, is much less amenable to this form of construction, because of its irregular shape (even with a fairing) and the need for several cutouts. Many machines described as monocoques should more properly

be said to have fabricated backbones. The original Honda NR500 racer was an exception, with the fairing an integral part of the bike. An unfortunate result, considering the frequent attention required in racing, was the fact that extensive dismantling was necessary for many routine maintenance tasks. (It is worth noting that the British Maxton concern was commissioned to produce a more conventional frame for this machine.)

A true candidate for the term monocoque was the NSU 'flying hammock' record breaker of the early 1950s. In this, the primary structural strength was provided by the fully streamlined shell, hand-beaten in 1 mm-thick high-tensile aluminium alloy and reinforced, to take care of the points of application of chassis loads, by transverse channel-section members riveted in place.

Structural engine

Potentially, this is the most efficient way to build a bike with a large engine—and the post-war Vincent V-twin was an outstanding example.

The principle here is to use the inherent stiffness of the engine-gearbox unit to provide the major support between the steering head and the rear-suspension pivot. If that pivot is incorporated in the rear of the gearbox casting, then a simple lightweight structure will usually suffice to joint the steering head to the top of the engine.

On the Vincent the rear-fork pivot was clamped between aluminium-alloy plates at the back of the gearbox, while both cylinder heads were attached to a fabricated backbone that doubled as an oil tank and incorporated the bolted-on steering head.

More recently, the Norton-Cosworth parallel-twin racer demonstrated the potential of this method. The rear-fork pivot was cast in the back of the gearbox while the steering head was supported by a triangulated frame bolted to the cylinder head and a simple structure reached rearward to support the seat. Unfortunately, development was halted by the Norton collapse.

Left Monocoque construction was initially used on the ill-fated Honda NR500 V-four grand-prix four-stroke. One of the problems was inaccessibility for maintenance and a tubular frame was substituted

Below The highly successful NSU flying-hammock record breaker of the mid-1950s had a true monocoque construction. The shell was in 1 mm-thick aluminium alloy reinforced by bulkheads

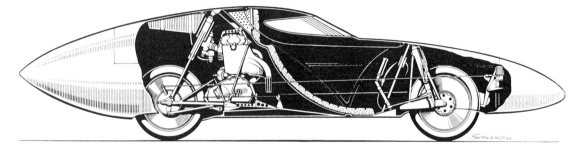

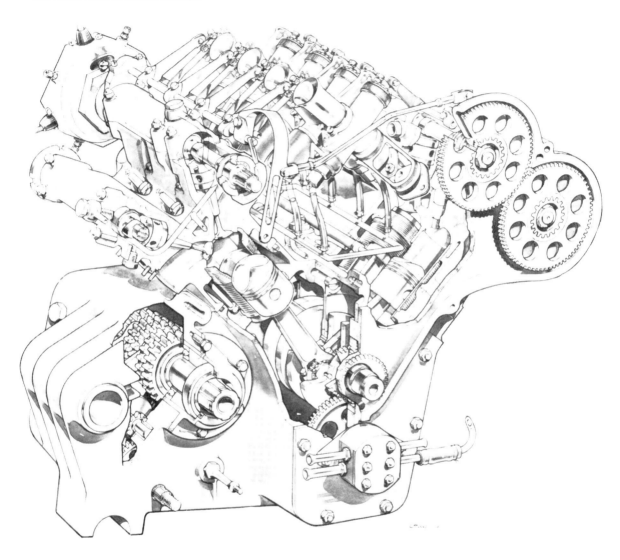

Stiffness was enhanced in the Moto Guzzi V8 racer and record breaker by incorporating the rear-fork pivot lug in the back of the gearbox casting (*MCW*)

Opposite, top Despite theoretically poor structural efficiency, multi-tubular frames such as this Nico Bakker example can give good results (Castricum)

Opposite, bottom With the engine installed, this Harris frame has a high degree of stiffness, helped by double-triangulation at the steering head, supported by eight tubes (Parker)

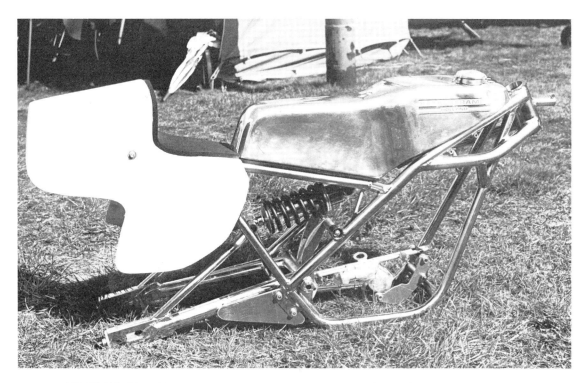

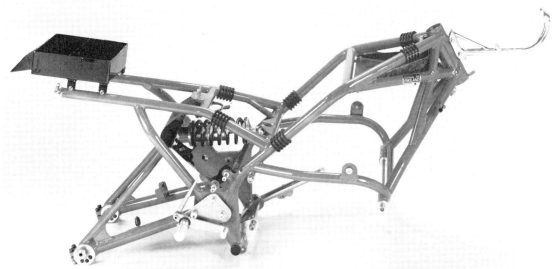

Conventional multi-tubular

Although they are the most common, these frames are potentially the worst in terms of structural efficiency, their layout being determined primarily by availability of space. They comprise medium-size tubes bent around the engine to connect the steering head to the rear-suspension pivot.

The tube diameter is too small to gain much from the bending and torsional stiffness of the sections themselves, as in the case of a back-bone frame. Moreover, the layout is rarely such as to provide any significant triangulation. Indeed, many of these frames are relatively flexible and acceptable road behaviour is obtained only through the structural effect of bolting in a rigid engine.

Despite these shortcomings, many machines with this type of chassis have achieved excellent handling (the featherbed Manx Norton being the most renowned), though only at the expense of weight, for a heavier frame is needed to give the required rigidity.

In this context, it must be borne in mind that frame stiffness is not the only factor influencing handling. For example, when Norton substituted a racing parallel-twin Dominator engine for the single-cylinder Manx unit in 1961 the handling suffered slightly as a result of the centre of gravity being raised—for the wider engine had to be installed higher to clear the bottom frame rails.

To summarize the structural considerations, the optimum chassis design depends on the size and shape of the engine and the intended

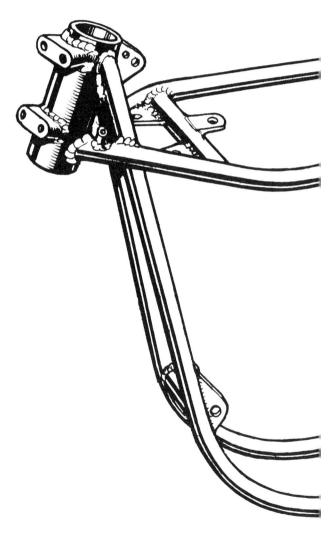

Right **Probably the most famous of all multi-tubular frames—the duplex-loop Norton featherbed. Best handling was achieved with the single-cylinder engine because of the low installation possible (*MCW*)**

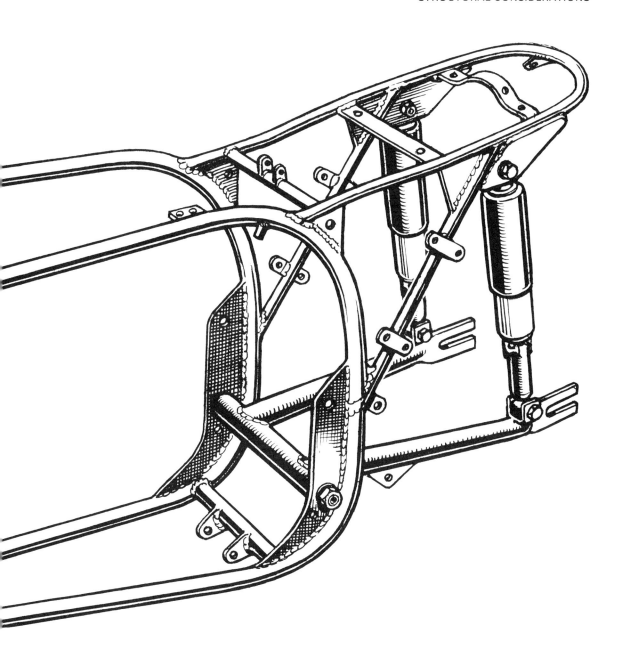

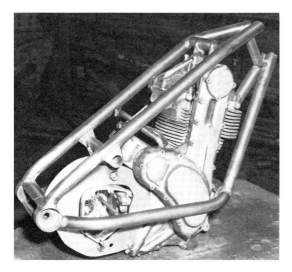

An interesting multi-tubular variant designed by the late Bob McIntyre for an AJS 7R and later campaigned by Jack Findlay with a Matchless G50 engine. Engine and gearbox are low slung and well supported (*MCW*)

purpose of the machine. While a small tubular or pressed-steel backbone suits a moped for both structural and economic reasons, large-capacity machines may best be designed with the engine as the main structural member. This method is rarely open to the specialist or low-volume producer, however, as the engine should really be specially designed for such use.

Where the engine is flat or compact, triangulation offers high structural efficiency, though with more complexity than a backbone. Both these approaches lend themselves to specials builders.

For mass-produced sports machines, tradition and market acceptability will probably ensure the survival of the multi-tubular frame for some time yet, perhaps with an increased use of steel pressings. In its favour is its ready adaptability to various engine sizes and styles. But it is difficult to mass-produce cheaply and that factor, rather than any technical one, may eventually woo manufacturers away from it to the structural-engine concept.

It is difficult to visualize much application for genuine monocoque structures.

Materials and properties

Before deciding which material is most suitable for any particular component, we clearly need to know something about material properties. The main properties of concern to us are:

- Strength
- Stiffness
- Density (or specific gravity)
- Ductility
- Fatigue resistance
- Available joining methods
- Cost of material
- Cost of machining and working

The relative importance of these properties depends on the purpose for which the machine is intended. For example, low cost is much more important for a mass-produced moped than it is for a works grand-prix racer, where cost is secondary to low weight.

In the previous chapter we defined the terms strength and stiffness in relation to a complete structure. The same concepts apply to a single piece of material and, when quantified, provide a yardstick for comparing different materials. The term stress—more particularly ultimate tensile stress (UTS)—is used as a measure of strength.

Stress is expressed as the force applied per unit of cross-sectional area; e.g. if we apply a load of 1000 lb to a piece of material of 1 in.-square section, then the stress in the material is 1000 lb per square inch (psi). The UTS is the stress under which the material breaks completely.

In some cases, other stresses may be a more valid criterion for comparison. Yield stress, for example, is the stress where permanent deformation begins and is useful when comparing ductile materials.

Under load, all materials deflect to some extent. This deflection is called strain and simply expresses the proportional change in dimensions; e.g. if we pull a 100 in. length of material until it stretches by 0·1 in., then the strain in the material is

$$\frac{0·1 \text{ in.}}{100 \text{ in.}} = 0·001.$$

As we saw in Chapter 4, stiffness is the ratio of the applied load to the deflection it causes. This is defined by Young's Modulus, which is simply the applied stress divided by the resulting strain; e.g. if a stress of 10,000 psi is needed to produce a strain of 0·001, then Young's Modulus for the material under stress

is $\quad \dfrac{10,000 \text{ psi}}{0·001} = 10,000,000 \text{ psi}$

$$(\text{i.e. } 10 \times 10^6 \text{ psi}).$$

Density is a measure of weight per unit volume; hence, size for size, it compares the weights of different materials. We get the same comparison from specific gravity, since that is just the density of any material compared with that of water.

Ductility determines the type of failure exhibited by a material. If it undergoes much permanent deformation before final fracture it is said to be ductile. But if it fails suddenly, with little prior distortion, then it is brittle.

Generally speaking, a ductile material is pre-

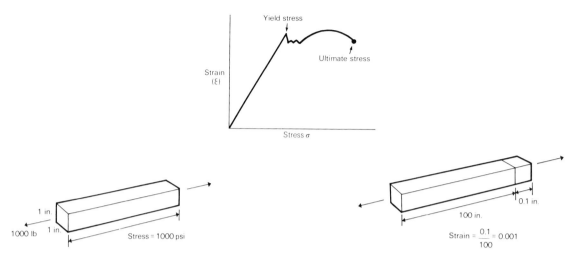

Fig. 5.1 **Typical stress-strain curve for steel; the initial slope of the line is the Young's Modulus**

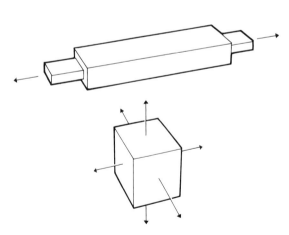

Fig. 5.2 **When a bar is stretched along only one axis, the longitudinal extension is accompanied by a lateral contraction. Triaxial stress prevents such contraction and the material may fail in a brittle manner**

ferable because it can withstand a certain amount of overloading without total failure. For example, if a cast wheel is made from a ductile material it may be only dented or buckled on hitting a kerb or sharp pot-hole; but if it is brittle a section of the rim may break off or the spokes may fracture.

Ductility can also be strongly influenced by the type of loading. An extremely ductile material (even rubber) can fail in a brittle manner if subjected to triaxial stress—i.e. to loading that results in a component of stress on each of three mutually perpendicular axes.

Unfortunately, as we increase the strength of materials we usually increase their brittleness too. So, the selection of a suitable grade of metal and its heat treatment is a compromise that can be determined only in the light of the component's duty and applied loads. As a rule, cast metals are less ductile than their forged or wrought counterparts.

As mentioned in the preceding chapter, fatigue characteristics vary with the type and condition of a metal. An additional point here is that a brittle metal is more likely to suffer a fatigue fracture than is a ductile one—the non-brittle material will distort in such a way as to reduce the stress concentrations and so lessen the possibility of ultimate failure.

Typical properties of some common materials

Material	Ultimate tensile strength tons/sq in	Specific gravity	Young's Modulus $\times 10^6$ psi	Specific stiffness
Grey cast iron	10–14	7·3	18	2·5
Structural steel	37–43	7·8	30	3·8
Aluminium alloys	19	2·7	10	3·7
Magnesium alloys	12	1·7	6·2	3·7
Titanium	62	4·4	17	3·8
Polythene mouldings	0·5–2·0	0·95	0·01–0·015	0·1
PVC	4·0	1·4	0·35–0·6	0·4
GRP	10–12	1·4		

The above properties are a rough guide only, as the tensile strength may vary considerably, depending on the metal composition or alloy and its state of heat treatment and working. The specific gravity and Young's Modulus do *not* vary in this way.

By specific stiffness we mean the ratio of Young's Modulus to specific gravity, which is a measure of stiffness per unit weight. For the metals in our table, except for cast iron, this ratio is almost constant at 3·7 to 3·8. Thus, while aluminium alloys are approximately only one-third as dense as steel, their stiffness too (size for size) is only one-third that of steel.

To put it another way, weight for weight, aluminium has approximately the same stiffness as steel, magnesium and titanium. However, this applies only to tension and compression stresses. Stiffness in bending and torsion depends not only on the modulus and cross-sectional area (proportional to weight) but also on the second moment of area (moment of inertia)—and this provides a clue to the efficient use of lightweight materials.

Consider a solid round bar subject to a bending load.

To determine the bar diameters that will give the same stiffness in both steel and aluminium, we need the product of E (Young's Modulus) and I (second moment of area) to remain constant. Now I depends on the fourth power of

bar diameter, while bar weight depends on diameter squared and material density. Calculation shows that—since aluminium has one-third the density and one-third the modulus of steel—the diameter of the aluminium bar needs to be larger by 32 per cent, at which its weight will be only 58 per cent of that of the steel bar.

In a motorcycle chassis, of course, we are seldom concerned with solid bars but mostly with round tubes. To maintain the same bending and torsional stiffness with a less dense tube material, we can either keep the diameter the same and increase the tube thickness or (more efficiently) keep the thickness the same and increase the diameter. Between the two extremes is a wide choice of proportions.

If we increase the tube diameter and use the same wall thickness, we find that an aluminium tube needs approximately twice the diameter of a steel one but weighs only 70 per cent as much.

Thus the most efficient way to use lightweight materials is to make the sections as large as possible consistent with maintaining a practical wall thickness. But, in maintaining similar structural characteristics to those of steel, our light alloy tube will weigh more than a simple comparison of density indicates. The density of aluminium is 33 per cent that of steel but the structural weights of our bar and tube in the foregoing examples are 58 and 70 per cent respectively.

The terms chrome-moly, T45 and 531 are frequently bandied about as though they have some magical significance, implying extra stiffness and lightness, to such steels.

In fact, these terms refer to steels with alloying elements calculated to enhance strength, particularly strength after welding. Their Young's modulus, hence stiffness, is no different from that of other steel alloys, nor is their density. Hence, if they are substituted for lower-strength steels and the same size tubing is used, then the weight and stiffness of the frame will be unchanged. Where they score is in the ultimate load that the frame will take before breaking.

If stiffness is no problem with a particular structure or member, then the use of high-strength tubing will permit thinner walls, hence reduced weight. But if stiffness is vital, then the best way to use this tubing is to reduce wall thickness and increase diameter. In this way only can stiffness be improved and weight reduced.

A Reynolds trade name, 531 is often referred to as chrome-moly, whereas it is actually a manganese-molybdenum steel, which Reynolds claim has superior properties to those of a chrome-molybdenum steel.

The alloying elements in Reynolds 531 are as follows:

Carbon	0·23 to 0·29 per cent
Silicon	0·15 to 0·35 per cent
Manganese	1·25 to 1·45 per cent
Molybdenum	0·15 to 0·25 per cent
Sulphur	0·45 per cent maximum
Phosphorous	0·45 per cent maximum

Its minimum strength properties are:

	As drawn	After brazing
Yield stress	45 tons/sq. in.	40 tons/sq. in.
Ultimate stress	50 tons/sq. in.	45 tons/sq. in.

These figures indicate the excellent retention of strength after brazing, which is a great boon. The use of this and other high-strength tubing is normally confined to competition machines; for roadsters the extra cost is not usually warranted.

Now that we have dealt with the principles underlying the selection of materials, let us consider the choices open for various components.

Frame

Steel is easily the most common material here, either as tube or sheet, depending on design.

There are several reasons for its choice but they generally boil down to cost, viz:

1 Raw material cost is relatively low.
2 Well developed manipulating and joining techniques are available.
3 Young's Modulus is high, so the required stiffness can be obtained with small tube sizes.

Aluminium has often been used for specials and racing machines in the form of mono-coques and large-section backbones such as the fabricated Ossa and Kawasaki mentioned earlier. Cast-aluminium backbones have been tried by Eric Offenstadt in France and Terry Shepherd in England.

More recently, tubular aluminium frames have appeared on works racers, with Yamaha taking the initiative. This trend started cautiously, when just the pivoted rear fork was made in light alloy (as it is on several roadsters now), before spreading to the complete chassis. In the development of aluminium frames, however, it is interesting to note that tube sizes have increased rapidly to compensate for the low Young's Modulus, as explained earlier. A great help in this context would be the spread of proper triangulation, with low-modulus materials.

It must be remembered that the fatigue characteristics of aluminium are such that failure is inevitable eventually in components subjected to alternating stress; hence limited life must be accepted. In the case of works racers, their natural rapid obsolescence makes this not a serious problem. But for touring machines, where long life dictates lower stress levels, it may well be that aluminium's weight advantage over steel is lost.

Tubular frames have also been made of titanium—as, for instance, on the BSAs for the world motocross championships in the 1960s. The likely drawback here was excessive flexure, because photographs suggest that the tube sizes were no larger than in the team's successful steel frames, whereas titanium has approximately half the modulus of steel.

This does not mean that titanium is unsuitable for frames, but rather that design must take account of the properties of the materials used. With its low weight and high strength, titanium is probably used to best advantage in a triangulated design. Its chief disadvantages are high cost and the sophisticated welding techniques required; but its corrosion resistance is excellent.

Magnesium, both cast and fabricated, has been used for backbone-type frames. Besides high cost and welding difficulties, however, it has the added disadvantage of limited life as a result of both fatigue and corrosion.

Porsche accepted this limitation in their 917 sports racing cars, the chassis of which comprised welded-up triangulated structures in magnesium tubing—so the technique might be worth trying on racing motorcycles, given ample financial resources.

Peter Williams was a pioneer of cast light-alloy wheels for racing. Here is Tony Foale's later version of the design, in which the sprocket and brake-disc carriers are cast integrally

The use of composite materials, such as carbon fibre and Kevlar reinforced plastics, is on the increase in Formula One car racing. The monocoque style of chassis construction there lends itself to this approach much more than does a motorcycle; but clever design and future development in materials may alter the picture, though at present construction costs are very high.

Wheels

For most of motorcycle history, the traditional wheel was a composite of hub, spokes and rim. Hubs have been made in steel, cast iron, aluminium and magnesium (the light-alloy

Traditional wire wheels were occasionally fitted to this Foale-framed 700/750 cc Yamaha four on which Steve Parrish won the 1976 British championship. Note also the drilled brake discs

hubs usually having a cast-iron brake drum). Spokes are of steel, usually with brass nipples, though these are sometimes in aluminium for racing. Rims have mostly been of steel, except that aluminium has taken over for sports and racing machines and some roadsters.

Since the late 1960s, however, cast wheels have become increasingly popular, first for racing (where magnesium predominates) then on roadsters, where cost and corrosion problems favour aluminium.

In magnesium, a properly designed cast ·wheel may well be lighter than a steel-spoked wheel with an aluminium rim and magnesium hub; but cast-aluminium wheels usually have a weight penalty though they may be stiffer laterally and run more accurately.

An ingenious sheet-aluminium design by Tony Dawson consists of left and right pressings riveted together at the rim and bolted to a cast hub. The higher strength of the wrought material used enabled these wheels to compete with cast magnesium for weight, while the greater ductility of the sheet material gives a high safety factor. It would be interesting to see this technique tried with magnesium sheet.

Considering the speed of development in plastics technology, it may soon be a practical proposition to make wheels in this material, thus considerably reducing unsprung mass. Such wheels are already well established on children's BMX cycle-type machines.

Fuel tank

Steel is the usual material here for roadsters, aluminium for racers. To prevent cracking, care must be taken to isolate aluminium tanks from vibration.

Plastic tanks—both glassfibre-reinforced and moulded thermoplastics (ABS or similar)—have been successfully used for competition duty (particularly on off-road machines) but the Construction and Use regulations forbid the use of non-metallic tanks on the public roads in Britain.

Brake discs

Stainless steel is most commonly used on road machines because its freedom from corrosion preserves a smart appearance. But it has the disadvantage of poor wet-weather performance, though this is improved by drilling the discs to break up the surface water film.

Above **A future motorcycle trend? This BMX bicycle has ultralight plastic wheels (Smith)**

Top **Tony Dawson's Astralite wheels consist of left and right sheet-aluminium pressings riveted at the rim and bolted to the cast hub. A similar construction was later adopted by Bimota**

131

Another improvement has been brought about by a change in brake-pad material.

Cast iron is functionally superior to stainless steel; consequently it is widely used on racing machines and some Italian sportsters. Its chief drawback is its poor appearance in damp weather, due to rust.

To reduce unsprung mass, many racing machines have aluminium discs with a hard surface coating. This may take the form of hard anodizing, chromium plating or, more expensively, a plasma-sprayed material. Generally, such discs have a shorter life than the cast-iron variety.

Because of the higher coefficient of expansion of aluminium alloys care must be taken to prevent distortion due to temperature differentials between the disc operating surface and the hub. The best method seems to be to use a flat disc, attached in a semi-floating manner to a central carrier. See page 103.

To allow for the greater radial expansion of the aluminium disc there must be more clearance between its periphery and the caliper, otherwise the brake could lock when hot. High disc expansion can also cause a hard surface coating to crack if it is not of a compatible material.

Bodywork

The use of steel or aluminium for seats, mudguards, fairings and suchlike has been largely superseded in racing by GRP. On road machines, too, metal is being ousted for these components in order to save weight; but in this case thermoplastic mouldings are commonly used, some of which have greater flexibility, which reduces the chance of permanent damage in a minor accident. A disadvantage, however, is their tendency to look tatty in time as a result of scratching and other surface blemishes.

Engine mounting

Simple though it may seem, the method of mounting the engine in the chassis calls for careful consideration if annoying, or even destructive, vibration is to be avoided. The basic requirement is to support the engine in such a way that all its loads are adequately dealt with. These loads are of three sorts: a) engine weight and inertia loads due to road shocks; b) unbalanced reciprocating forces in the engine: c) final-drive chain tension, which can be extremely high in a large-capacity machine in a low gear.

In the traditional old singles and 360-degree vertical twins (pistons in step) vibration was a serious problem, exacerbated in many cases by relatively flexible frames that could very well vibrate in sympathy with the engine. If a frame resonance coincided with a particular engine speed, then the vibration transmitted to both rider and structure would be magnified.

A common method of reducing this vibration (which, in a big single, was aggravated by pulsating torque reaction) was to fit a cylinder-head steady (i.e. a stay or bracket bracing the head to the frame). This worked by using the bulk of the engine to stiffen the frame, so raising its resonant frequency and reducing the magnitude of the vibration. In some cases, however, fitting a head steady made the problem worse.

Another important influence on the level of vibration, especially in a big single or parallel twin, is the engine's balance factor—i.e. the proportion of the reciprocating mass that is counterbalanced in the flywheel assembly.

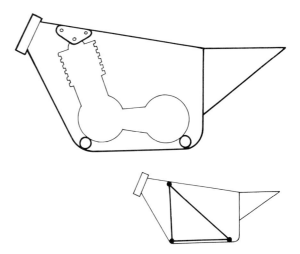

Fig. 6.1 **A cylinder-head steady fitted as shown (top) effectively triangulates the frame (bottom), so raising its resonant frequency**

This factor determines both the direction and magnitude of the *primary* out-of-balance forces.

With a zero balance factor these high forces act in line with the cylinder axis. By using a 100 per cent balance factor, the direction of the forces is turned through 90 degrees (i.e. fore and aft for a vertical cylinder). For intermediate balance factors, the forces can be reduced in magnitude while their direction varies between the two extremes.

At first sight, it seems best to use a balance factor of 50 per cent and so minimize the out-of-balance force, in whichever plane—as, for example, did AJS and Matchless in their 500 cc parallel twins. In practice, however, frames differ in their stiffness in various directions and the best balance factor can be determined only by experiment.

At a time when proprietary engines (such as Blackburne, JAP and Python) were installed in different makers' frames, it was normal to alter the factor to suit each particular frame. Indeed, adapting any make of engine to a frame for which it was not intended became a specialized art.

For various reasons, the problem of vibration has diminished with time. In the first place, large singles nowadays have contra-rotating balance shafts to oppose the out-of-balance forces; anyway, the single-cylinder engine is mainly confined to small-capacity machines and is predominantly of two-stroke type, which has lighter reciprocating masses. Secondly, some parallel-twin four-strokes adopted 180-degree spacing of the crankpins (as in a two-stroke) so cancelling the primary inertia forces, albeit at the cost of a rocking couple; others kept the pistons in step (360-degree cranks) while incorporating balance shafts. Finally, engines with more than two cylinders are inherently smoother anyway.

Nevertheless, some high-revving fours, particularly two-strokes, still generate enough high-frequency vibration to fracture components. Rubber mounting the engine can help here, although the solution may be anything but straightforward. For example, the 125 cc Suzuki grand-prix twins of the 1960s were said to have shattered the insulators of their sparking plugs as soon as the engines were rubber mounted.

Clearly, the simplest form of flexible mounting is to fit bonded-rubber bushes in the engine's attachment lugs. With final chain drive, however, there is the disadvantage that the large forces in the chain pull the engine back in the bushes, so reducing their effective-

ness and increasing chain slack. On the other hand this builds some cushioning into the drive.

With their Isolastic system, Norton skirted round the chain-pull problem in the Commando by coupling the engine, gearbox and rear fork rigidly together and rubber mounting the whole assembly in the frame. Thus the rear suspension was flexibly joined to the front. To minimize lateral compliance, a shim adjustment was incorporated; and, provided this was correctly set, handling was satisfactory. However, the chore of periodically making the adjustment was often neglected to the detriment of handling.

A better solution was James Love's patented Vibratek system described (in a Triumph Bonneville application) in *Motor Cycle Weekly* for 15 September 1979. The idea was to reduce the engine's balance factor to zero (so that the primary inertia forces were confined to the vertical), then pivot the engine at a precisely calculated point well to the rear, allowing a tiny vertical movement between rubber stops at the front. Because the main unbalanced forces were vertical, little of these were transferred to the frame via the pivot, the engine unit tending only to rotate about this point.

The location of the pivot point was crucial and depended chiefly on the relative positions of the engine's centre of gravity and its crankshaft. Unfortunately, in this case the required

Fig. 6.2 **Principle behind Vibratek system**

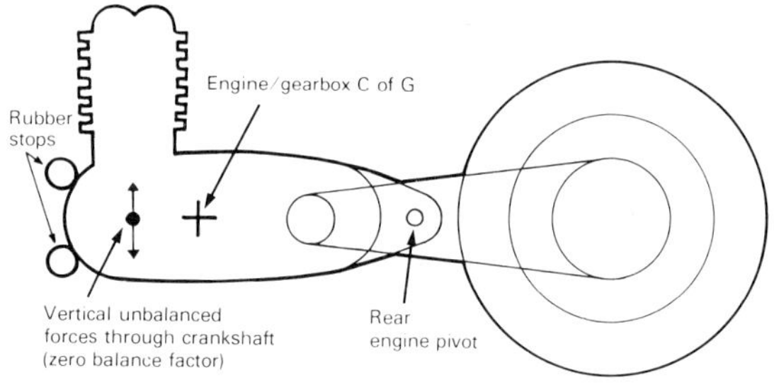

Rubber stops

Engine/gearbox C of G

Vertical unbalanced forces through crankshaft (zero balance factor)

Rear engine pivot

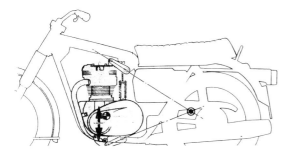

In James Love's Vibratek conversion of a Triumph Bonneville, the theoretical engine pivot point proved impractical; hence the two links giving virtually the same effect

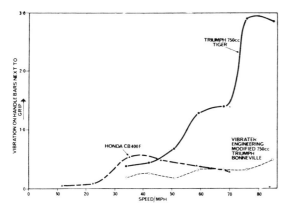

Vibrations induced in handlebars of modified Bonneville are on average 25 per cent of those in handlebars of Tiger and over much of the range less than those induced in the handlebars of a 4 cylinder machine of just more than half the capacity

pivot point proved to be within the rear-wheel area, which was clearly impractical. But, since the engine's angular movement about the pivot was extremely small, virtually the same movement was obtained by attaching the engine to the frame by short links top and bot-

Fig. 6.3 **A similar approach by MZ on a 250 cc single-cylinder roadster**

tom, so arranged that if they were extended rearward they would meet at the theoretical pivot point.

Tests on a rolling road indicated an enormous reduction in the vibration felt by the rider.

A somewhat similar system has been used for many years on 250 cc MZ roadsters and, though not perfect, worked quite well. As the drawing shows, the engine was pivoted on nylon bushes on the rear-fork spindle and rubber mounted at the back of the cylinder head.

The engine's balance factor is not known to the authors but the layout results in a fore-and-aft component of the inertia forces being fed into the frame through the pivot and rubber mounting bushes at lower engine speeds, although most of the loads at higher rpm would be reacted against the engine mass through its centre of gravity if the balance factor was zero.

Naturally, a rubber-mounted engine cannot be used in a structural way. In most multi-tubular frames a rigidly mounted engine stiffens the chassis considerably; hence the induced forces must be taken into account when designing the frame brackets and engine lugs involved.

Where a chassis is sufficiently rigid on its own account (e.g. a good triangulated or backbone design) it is best not to attach the

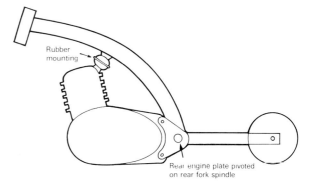

Rubber mounting

Rear engine plate pivoted on rear fork spindle

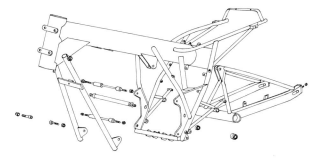

Left Designed to accommodate a variety of Japanese tall transverse fours, this Foale backbone frame has two tension struts to support the engine weight. The rear gusset plates provide mountings for the rear of the engine and the fork pivot (Sansum)

Below In the Norton Cosworth Challenge parallel twin the engine was the main structural chassis member, with the rear-fork pivot lug cast in the gearbox shell and a simple structure at the front to carry the steering head

Right The cylinder-head brackets used to adapt this Kawasaki engine for main structural use are spigoted into the head and so relieve the extended studs of fore-and-aft loads. Spigots 'A' fit into the counter bored head stud holes 'B'

Below A simple triangulated structure is bolted to the above brackets and the standard front mounting lugs on the crankcase. Specially made side castings replace the original parts to give support for the rear fork pivot bearings, which are coaxial with the gearbox sprocket.

Using the engine as the main structure on this Foale special resulted in a very rigid and light machine

engine at too many points. At the front, no more than one or two bolted-on tension struts may be needed to support the weight of the engine, while the chain pull may be taken through the rear engine mountings, preferably integrated with the rear-suspension pivot.

Benefits inherent in the use of bolted-on tension stays at the front include: 1) easy installation and removal of the engine; 2) elimination of bending moments at the upper end of the stays reduces the chance of failure; 3) ease of accommodating dimensional tolerances in manufacture.

If the engine is to serve as the main structural member, the rear-suspension pivot is best incorporated in the crankcase/gearbox castings. It then remains only to build a simple structure at the front to support the steering head, leaving a minimum need for additional framework elsewhere to support the tank, seat and rear suspension. The Norton Cosworth Challenge was designed in this way.

Naturally, this scheme is feasible only if the engine has been suitably designed in the first place. Many of the large multi-cylinder engines that are sufficiently substantial lack suitable mounting lugs. In these cases, the rear-suspension pivot may be housed in plates rigidly bolted to the rear engine-mounting lugs, while through bolts from crankcase to cylinder head may sometimes be lengthened to provide a suitable attachment for the front structure.

Provided the engine is inherently fairly smooth (e.g. a good four or V-twin) vibration is no great problem when it is used as the main structure. Since the additional substructures are usually short and rigid, their resonant frequencies are beyond the range of engine rpm.

This Honda CBX engine already incorporated cylinder-head lugs suitable for the steering-head structure and for the seat and shock-absorber anchorage

Practical frame building

This chapter is intended as a guide for those contemplating the construction of a one-off special or a relatively small production run. The techniques used in large-scale production may differ considerably from those described here, because of the much larger investment available to a large manufacturer for automation. The differences chiefly apply to processes such as welding, tube manipulation (bending and end-profiling), bracket and sheet-metal pressing and jigging.

The money spent on production facilities must be balanced against the extra construction time involved when using less-expensive equipment; and the final decision depends on the number of similar frames to be produced.

For low-quantity frame production, we are usually restricted to the following choice of methods and materials:

a) welded steel or aluminium tubing
b) welded steel, aluminium or stainless-steel fabricated backbones (monocoques)
c) riveted or cast-aluminium backbones, e.g. Offenstadt and Seeley 500 Suzuki shown opposite. (This type is highly specialized and will not be discussed specifically.)

Welding
Two basic methods are available—electric and gas—each subdivided into a variety of techniques, which will now be described, together with their pros and cons.

1) Electric
Arc (stick electrodes). In common use on construction sites and cheap home-welding kits

Above The frame for this 500 cc Seeley Suzuki was constructed from bonded and rivetted aluminium sheet

Top Offenstadt frame for Yamaha 350 cc TZ engine had a cast-aluminium backbone and rear fork. Note inside-out telescopic forks

for do-it-yourself purposes, this is probably the least suitable technique for frame construction, although satisfactory results may be achieved provided the tubing is relatively thick-walled, say, 2 mm or more.

It is true that the roadgoing versions of the famous Norton featherbed frame were welded in this way; but welding technology has advanced considerably in the intervening decades and it is likely that, if this frame was making its debut today, the following method would be used.

MIG (Metal Inert Gas), often called CO_2 welding. Here the weld filler material takes the form of a wire, which is continuously fed into the joint by the welding machine and is shrouded by a flow of inert gas. This may be CO_2 or a mixture of argon and CO_2 (usually in the proportions of 80:20 or 95:5 respectively) and its purpose is to prevent oxidization of the hot welded material by blowing away the air.

Nearly all production frames in welded tubing are nowadays produced by this process, and it is easily automated. However, this is not to say that it is unsuitable for manual use on low production runs or one-offs. Although it is usually associated with steel tubing, it may also be used for aluminium and stainless steel.

The advantages include:

a) speed, which reduces not only labour costs but also distortion, since the total heat input is relatively low
b) clean welds—no flux to be cleaned off or to adulterate the weld
c) tolerance to operator skill—sound welds can be achieved with less experience than is required with some other methods
d) gap filling is good, hence less time need be spent on joint preparation.

Disadvantages may be:

a) weld fillet is convex, so may cause undue stress concentration in thin-wall tubing. This is less likely to be troublesome in triangulated or backbone frames than in the typical bent-tube type; but if used on 18-gauge (1·2 mm) or thicker tubing, it should cause few problems.

b) the high capital cost of the equipment may not be justified for the enthusiast wishing to experiment.

TIG (Tungsten Inert Gas), also known as Argon-arc and, in America, as Heli-arc. In this system, an arc is struck between a tungsten electrode and the workpiece to provide the necessary heat, while the filler rod is fed in by hand. As in the MIG process, a gas shield keeps out the oxidizing air; however, in this case the gas is almost pure argon.

This system can produce welds of higher metallurgical quality than the other methods described and may be thought of as a clean electrical version of an oxy-acetylene torch. In experienced hands it is very versatile, since it can be used for welding steel, stainless steel, aluminium, titanium and even the inflammable magnesium.

Since this process takes longer than MIG welding, however, distortion may be greater; but extremely neat concave welds are possible, thus reducing stress concentrations. It is suitable for all thicknesses of tubing and excellent for sheet-metal applications such as fuel tanks, two-stroke exhaust boxes and stainless-steel or aluminium fabricated backbones.

Because the results can be of such high quality, this method is extensively used in the aircraft industry. Unfortunately the capital cost of the plant limits its use to those with sufficient throughput of work to warrant the expense.

2) Gas (usually oxygen and acetylene)

Fusion welding Here the flame is used to melt the parent metal while filler rod of similar composition is fed in by hand, as in the TIG system. Gas fusion welding is seldom used in frame building but is popular for the construction of fuel tanks (in both steel and aluminium) and exhaust systems, especially two-stroke expansion chambers. Weld quality, however, is inferior to that obtained by the TIG method but initial financial outlay is small.

Bronze welding (often incorrectly referred to as brazing)

In the construction of special frames this is by far the most widely used method. It is suitable for tubular steel structures, where one of its chief advantages stems from the lower temperatures involved. In all methods of welding, the heat reduces the strength of the parent metal in the vicinity of the weld, especially in the case of some of the higher-strength steels and aluminium alloys. In some applications, the full strength may be restored by subsequent heat treatment; but this is hardly practicable with a motorcycle frame and would probably lead to distortion in any case.

In bronze welding, however, the parent metal is heated to a temperature which, though high enough to melt the bronze filler, is well below its own melting point; the parent metal therefore retains much more of its original strength, which is one of its chief advantages. Various types of bronze are available, with different strengths and melting points, to suit different applications.

With care, the weld fillet can be concave and broad based to give a smooth change of section and minimum stress concentration.

For all these reasons, bronze welding is highly recommended by Reynolds—makers of the famous 531 manganese-molybdenum tubing, especially for thin wall sections.

Although it is easy for an unskilled welder to make a joint by this method, considerable skill is required for a quality job. Temperature is critical—too cold and adhesion is poor, the joint weak; too hot and some of the elements in the bronze may vaporize off, while the filler material may penetrate deep into the grain structure of the parent metal, leaving it weak and brittle.

While many successful motorcycle frames have been welded by this method, it is this possibility of embrittlement that accounts for bronze welding not being approved for joints in the primary structure of aircraft.

It is absolutely essential, when constructing a frame in this way, that no load is placed on the joint until it has cooled down completely—otherwise intergranular penetration will occur and subsequent failure will be almost inevitable.

Distortion

The very process of welding causes distortion, so corrective measures must be taken to ensure accurate frame alignment. The causes of distortion are rooted in the cooling process of the weld; and proper welding sequence is as important to the finished product as is the basic design of the frame.

Consider the simple weld shown in figure 7.1. While the weld material is still molten, the welded pieces remain at 90 degrees to one another. As the weld cools, however, after solidifying, the material contracts (at a rate determined by its coefficient of expansion) and tends to pull the joined components to the position shown.

The linear contraction of the weld is, of course, only small but the movement at the end of the members is greatly multiplied by the leverage effect. That is why weld materials with a low expansion rate cause fewer problems through distortion. Bronze, with its high coefficient of expansion, is at a disadvantage here.

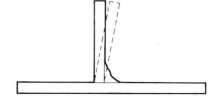

Fig. 7.1 **When a weld cools it contracts, causing distortion of the structure**

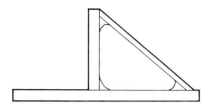

Fig. 7.2 **The added member prevents distortion of the joint on cooling but introduces residual stresses in the welds**

We can counter the effects of this type of distortion by putting other members in the structure, as shown in figure 7.2. Unfortunately, this has the disadvantage of building in stress at the weld zones; for, as the weld cools it is prevented from contracting and thus becomes subject to tensile loads. These built-in stresses add to those induced by operational loads and so tend to weaken the chassis.

With a welded frame it seems, then, that we have to choose between distortion and built-in stress. To a large extent this is true but several techniques are available to alleviate the situation. If, in our original example, we tilted the vertical member to the left before welding, then the subsequent contraction would bring it back to 90 degrees—giving no distortion and minimum stress.

If we consider the functional aspects of a chassis we see that fine tolerances are required in only a few instances, such as steering head to rear pivot housing, and the alignment of the final-drive sprockets. It is these relationships that are important; the path and location of the connecting structure are less critical.

We can make use of this fact in designing our welding sequence. If we complete the welding of the connecting structure before we attach the steering head and/or rear pivot, then most of the possible distortion will have taken place before these components are added. Furthermore, if each joint is completed and

allowed to cool before the next one is tackled, and the individual tubes are not jigged too rigidly, then there will be a minimum of built-in stress.

The alternative method of tacking all joints first, then finish welding, may result in the more accurate positioning of individual tubes; but the welding stresses will be higher, particularly in the case of bronze welding.

The following practical example shows just how the welding sequence can affect the final result. Originally—as shown in the first sketch—the backbone was welded to the steering head first, after which the gusset was added. On removing the frame from the jig, it was found that the steering-head angle had steepened.

Investigation showed that the welding on the gusset had caused the backbone to bend slightly. The solution was simply to weld the gusset to the backbone first, so allowing the distortion to take place before the steering head was attached.

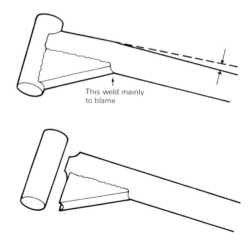

This weld mainly
to blame

Fig. 7.3 **The importance of welding sequence. Adding the gusset last (top) steepens the steering head on cooling. The solution was to attach the steering head last (bottom)**

Gussets

These are used at corner jonts to stiffen the structure, spread loads and/or provide mounting points. Their design must be given careful consideration if excessive stress concentration and consequent weakening of the frame are to be avoided.

We have already explained the benefit of attaching the gussets along the tube's neutral axis but there are additional methods of improving the assembly. For example, it is better to taper the ends of the gusset rather than cut them off abruptly. Here, ease of welding and manufacture conflict with the requirements of minimizing stress concentration.

Sometimes, where similar gussets are used on both sides of a tube, they are formed by one folded pressing, so that the two sides are joined by a web. In this case it is better not to complete the weld round the tube (i.e. at the ends of the web) as this would introduce more stress concentration.

The way in which a gusset stiffens a structure is not always understood. Imagine a tube attached to a rigid member and subjected to a bending force, as shown in figure 7.5. If there is no gusset to support it, the tube will bend over its entire length whereas, with a gusset, the bending is virtually restricted to the unsupported portion.

The amount of flexure is proportional to the cube (third power) of the unsupported length, so reducing that by a half increases the stiffness by a factor of *eight*. Hence even small gussets can stiffen a multi-tube frame considerably.

However, where a tube is stressed only in tension or compression a gusset can have no more than a minor effect. For that reason, they are seldom found on well-designed triangulated structures, except perhaps to provide mountings.

Jigging

A jig is simply a structure to hold the component parts of our frame in their correct relationship during welding. It may be simple or complex, depending on the quantity and quality of the frames to be produced.

For one-off jobs or experimental development work, the most versatile system is to use a *rigid* flat surface as a base to support easily made sub-jigs. An old flat lathe bed is ideal for this purpose, as squaring off the sides is easy and the centre gap is useful for bolting on fixtures.

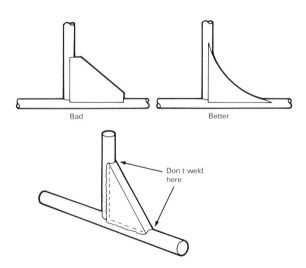

Bad   Better

Don t weld here

Fig. 7.4 **Stress concentration is lowered if gussets are tapered rather than sharply cut off. A folded gusset should not be welded at the ends of the web**

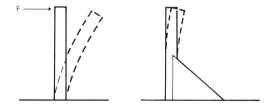

F

Fig. 7.5 **A gusset restricts bending mainly to the unsupported length of the tube**

Although all frame builders have their own preferences in the type of jig to be used, there are some common requirements, the most important of which are accuracy and rigidity. Nowhere is accuracy more vital than in the relative positioning of the steering head and rear-suspension pivot. And if the jig is insufficiently rigid, then welding stresses and/or the weight of the frame may cause distortion and make a mockery of the original accuracy.

A final tip on jigging. It is sometimes easier not to jig a frame in the same attitude that it will adopt on the road, even though you may find it easier to visualize the placing of the various components; otherwise it may require more complex jigging, including a long, high steering-head fixture, particularly when a flat bed is used as a base. If the frame is properly drawn first (say to $\frac{1}{4}$ scale) then simpler jigging may be possible, with the steering head clamped directly to the bed. It may be worth

Above **This jig was used to make the type of leading link fork shown on page 80. Note that the end supports allow the jig to be rotated for easy weld access. The support at the RH end incorporates a clamp to hold the jig in the position required. The block underneath is a counterweight to balance the jig. Show to the left is a typical MIG welder**

Opposite, top **If the frame is jigged in the attitude it assumes in service it may require relatively complex jigging, including a long, high fixture for the steering head. Note the use of a flat lathe bed for the jig base**

Opposite, bottom **If the same frame is rigged in this fashion then the steering head can easily be clamped to the jig at 90 degrees, resulting in a simpler, more rigid arrangement**

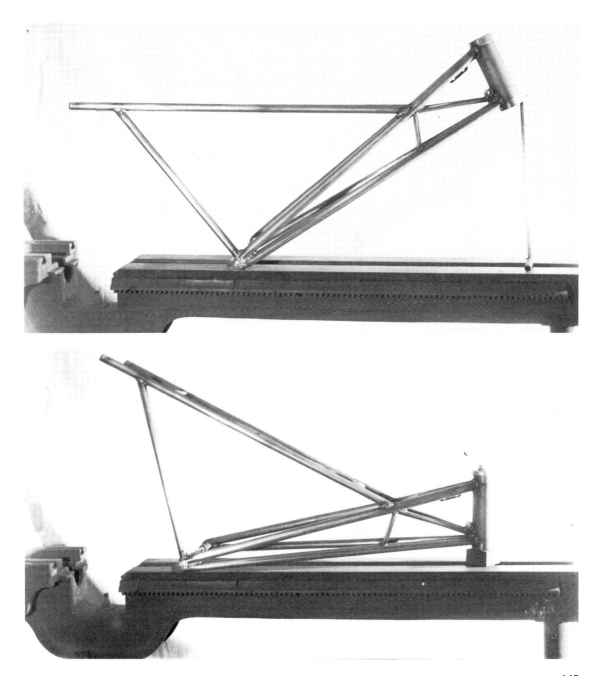

while making the jig rotatable, as this gives full access to welded joints without contortions by the operator and so helps improve weld quality.

Tube profiling

Except in the case of square or rectangular tubing, it is necessary to shape the tube ends to match the mating structure. This is often done with a milling cutter of appropriate diameter set at the required angle.

While this method presents no problem for a large production run, it is seldom that a specials builder has access to the necessary facilities. In most cases, however, two straight saw cuts at the end of a round tube will produce a nicely fitting joint.

Tube types

Round-section tubing has long been the most common shape in frame production—and not just because it is cheaper than other types. It is also the best section for resisting torsional and compressive loads and is equally capable of resisting bending loads in whatever direction.

However, other sections are sometimes used, mainly oval, square and rectangular, though occasionally we find tubing of taper section, as in the pivoted rear fork of some Velocettes. BMW have even gone to the expense of tapered oval tubing to match the tube properties to the applied loads, which is the chief justification for non-circular sections.

In Chapter 4 it was shown that a tube's resistance to bending is a function of the second moment of area; this depends on the tube shape and, except for a round section, differs about the neutral axis according to direction.

In each case of figure 7.7 the second moment of area is greater about the XOX_1 axis than about the YOY_1 axis; hence the resistance to bending is also greater about the XOX_1 axis.

So, if we know that the bending loads to which a frame member is subjected are mainly in one direction, we can tailor the tubing shape to achieve the most efficient structure. A typical example is a pivoted rear fork, where any sideways load produces different bending moments about the neutral axes of the fork arms. See figure 7.8.

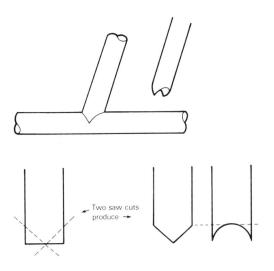

Fig. 7.6 **Simple tube profiling by two straight saw cuts**

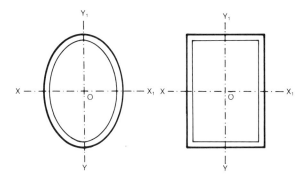

Fig. 7.7 **In oval and rectangular tubing, bending stiffness is greater about axis XOX_1 than YOY_1**

In this fairly typical layout—13 in. wheel radius and 10 in. between fork arms—the vertical bending forces due to a sideways load are $\frac{1\cdot3}{0\cdot5}$ i.e. 2·6 times the horizontal forces.

Among the specialist manufacturers a popular tube for this duty is 2×1 in. $\times 16$ g (50×25 mm $\times 1\cdot5$ mm) rectangular section. The relative second moments of area for this are 0·186 in.[4] and 0·0622 in.[4]—i.e. a ratio of 2·99:1, which closely matches the load ratio of 2·6:1.

But whereas the bending *load* acts at the wheel end of the fork, the maximum bending *moment* is at the pivot end. Thus, if our fork is to be structurally balanced we should follow Velocette's lead and use taper tubing, though the extra expense is seldom thought to be warranted.

The use of square-section tubing (which is gaining ground in aluminium racing frames) is more difficult to understand and probably owes much to fashion. If we consider a 1 in.-square tube of 0·0625 in. wall thickness (approximately 25×25 mm $\times 1\cdot5$ mm) then this has a second moment of area of 0·0346 in.[4] about any of its principal neutral axes.

Now if we substitute a round tube of equal *weight* and wall thickness we need a diameter of approximately 1·25 in.; and the second moment of area of that is 0·04123 in.[4]—i.e., 1·2 times that of the square tube. So, even on its most favourable axes, the square tube is less stiff in bending besides being less capable of handling torsional and compressive loads.

Tube sizes

Tube diameter and wall thickness are determined by the size, weight and power of the machine, also by the design family of the structure—e.g. triangulated, backbone, etc. Consequently the following is intended only as a general guide.

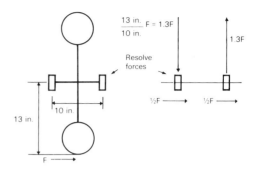

Fig. 7.8 In this pivoted rear fork, the effect of a side force at the tyre contact patch is resolved into horizontal and vertical components on the fork arms

Triangulated (space) frames

As mentioned earlier, the stresses in the members of this type of frame are mainly tensile and compressive; hence large diameter is not necessary to provide adequate bending and torsional stiffness. More important is the cross-sectional area of the metal. Thus a 0·5 in.-diameter tube of 2 mm wall thickness is interchangeable in most cases with a 1 in.-diameter tube of 1 mm wall thickness. However, if a member is long in relation to its size then it may buckle under compression before the ultimate stress levels are reached. In that case the larger diameter tube is preferable. Also, resonant vibrations create more of a problem with long, thin tubes.

Typical sizes for steel tubing may range from $\frac{1}{2}$ in. diameter $\times 20$ to 16 swg to 1 in. diameter $\times 20$ to 18 swg. For the longest tubes needed on a motorcycle it is unlikely that diameters of more than 1 in. would be needed. More commonly, we would expect to use tubes of, say, $\frac{3}{4}$ and $\frac{7}{8}$ in. diameter.

Bent multi-tubular frames

Since this chassis type gets its stiffness from a mixture of triangulation and the inherent

bending and torsional stiffness of the tubes themselves, it is more difficult to recommend suitable sizes; these depend more on individual design.

For the main structure, sizes may range from $\frac{7}{8}$ in. diameter $\times$ 16 to 14 swg, for smaller machines, to 1·5 in. diameter $\times$ 18 to 14 swg, for larger ones.

Where the subframe supporting the seat is not intended to contribute to the main structural stiffness, it may be made from smaller tubing—e.g. $\frac{5}{8}$ or $\frac{3}{4}$ in. diameter $\times$ 18 to 16 swg.

Tubular backbone frames

Even on smaller machines, the main member is unlikely to be stiff enough if much under 2 in. diameter $\times$ 18 to 16 swg. On larger machines, though, there is little to be gained by going above some 3 in. diameter $\times$ 18 to 14 swg. With this diameter it is better to avoid excessively thin walls, otherwise cracking may result at joints as a result of local buckling.

When designing this type of frame, great care must be taken to feed in the loads in such a way as to minimize the risk of such local buckling. This can be done by spreading the loads over a wide area, welding gussets and plates along the neutral axis wherever possible, and sometimes incorporating local reinforcement as shown in figure 7.9.

Fabricated backbones

Depending on detail design, 22 or 20 swg sheeting may be used here, with possible thickening in regions of high stress or load application.

This type of frame derives its stiffness from its large cross-section; overall shape is not

usually critical. Once more, however, the need to avoid local buckling cannot be over-emphasized and considerable thought must be given to all connection points if a satisfactory life is to be expected.

Frame finishes

Once we have constructed our frame we usually need to give it some coating to prevent corrosion and enhance appearance. The finishes normally available include:

1) Plating—chromium or nickel
2) Painting—a wide variety, including stove enamelling and epoxy powder
3) Plastic coatings
4) Anodizing—for aluminium parts

Plating

Although attractive in appearance, this can be expensive if a first-class result is required. It also tends to highlight any visual flaws, such as lumpy welds and scratches.

Acids from the plating process can become trapped in some tubes, either if the joints are not fully sealed by welding or if drain holes are not provided. By causing stress concentration, such holes can lead to premature fatigue failure. Trapped acids can eventually give rise to internal corrosion and can leach out at the welds, so causing rust areas and spoiling the finish.

A further risk with plating is hydrogen embrittlement. Here, hydrogen evolved during the plating process is trapped in the grain

Fig. 7.9 Detail design is very important where loads are fed into thin-walled large-diameter tubes. Here the insertion of a cross tube prevents local buckling

boundaries of the steel and can lead to failure. In this respect nickel plating is preferable to chromium. Indeed, in Formula 1 car racing chromium plating is banned on suspension parts for safety.

Despite the potential pitfalls, it has to be said that Rickman Metisse frames, among others, have long been nickel plated, seemingly without trouble.

Painting

New types of paint become available almost daily, hence advice should always be sought from the manufacturers.

Paints such as cellulose may be satisfactory for tanks and other bodywork but do not match up to the needs of the frame, where traditional stove enamelling produces the best all-round finish. The more modern electrostatically applied epoxy powders have their devotees but, though quite resistant to damage, are less amenable to touching up.

Plastic coating

A relatively new process for frames, this gives an excellent finish when first applied, since it conceals flaws, but is less impressive in the long term because scratches cannot be so easily polished out or touched up as they can in paint. Furthermore, if damage should penetrate to the underlying metal, moisture may spread for a considerable distance under the adjacent coating, causing widespread corrosion and lifting the plastic. With paint, on the other hand, any corrosion is local to the damage and easily repaired.

Anodizing

Although some aluminium alloys have good corrosion resistance, the tell-tale white powder on aluminium parts is an all-too-familiar sight, especially when a bike is ridden on salted winter roads. Anodizing, which involves immersion in an acid bath, is a protective process that prevents such corrosion by putting a tough oxide film on the surface. This oxide layer may be dyed—usually, grey, gold, red, blue or black—to provide an attractive appearance as well as protection. Some alloys benefit more than others from anodizing and wrought material usually responds better than cast.

Design and development

Throughout the history of motorcycle development there have been many brave attempts to advance the state of the art through radical changes in design. Yet it needs only a cursory glance at contemporary production machines to realize that these efforts have largely been in vain. In overall concept, today's bikes remain similar to those of yester-year, differing mainly in evolutionary improvement.

Compare, for example, the featherbed Manx Norton of the early 1950s with the average present-day racer. Both have bent multi-tube frames of similar basic design; both have hydraulically damped telescopic front forks mounted on a high steering head; both have pivoted-fork rear suspension and both carry their considerable petrol load in the highest possible place—above the engine. The 'monkey-on-a-stick' riding position is also unchanged.

There have, of course, been improvements during the several decades since the advent of the featherbed Norton—but they have been detailed rather than fundamental. For example, tyres have improved considerably in grip (in size, too, though that's a mixed blessing) to the benefit of roadholding, cornering and braking. The brakes themselves have changed the most, with hydraulically operated discs superseding cable and rod operated drums on all high-performance machines. In many cases, suspension characteristics have been refined, the most significant trends being towards pressurized, gas-filled shock absorbers, externally adjustable damping and triangulated rear forks.

There is, of course, little point in embarking on a new design unless it is an improvement on the one it replaces—and improvement means different things in different markets (e.g. speed for racing, economy for commuting). But why have improvements come only through evolutionary changes and not through revolutionary new ideas? It seems inconceivable that the basic layout of the earliest motorcycles was so intrinsically excellent that fundamental improvement was impossible—much more likely that radical advances in design have been killed by commercial imperatives, even on competition machines.

The major manufacturers are obliged to maximize the financial return on their activities—hence, in the heyday of the British industry, the entrenched reluctance to alter any design in such a way as to require substantial investment in new tooling. When rear suspension became popular (following racing successes) most manufacturers simply altered the rear end of the existing rigid frame so as to obviate rejigging the front. Another example was the use of a common frame for engines of different capacities and types (singles and twins). It was hardly surprising that progress was slow and the industry foundered.

Japanese manufacturers cannot be accused of reluctance to retool for new models. Indeed, they have moved too far in the opposite direction, producing a rapid proliferation of models with few common parts, to the consternation of importers and spares stockists. But even in Japan genuine advances (as opposed to gim-

micks) have been slow to appear, for the motorcycle market is essentially conservative and fashion-conscious, hence resistant to radical changes in design.

The surest recipe for rapid acceptance of change is success in racing. But even in this intensely competitive branch of the sport radical advance comes slowly. One reason for this is the domination of racing by the major manufacturers, who prefer their track machines to bear at least a superficial resemblance to their catalogue roadsters.

Paradoxically, another factor is the high ambition of the top riders, who are not willing to risk being uncompetitive during the development period of an unconventional machine, whatever its ultimate potential. Consequently, any such machine would have to be allocated to riders of lower status and—in competition with star riders on heavily sponsored factory bikes—would seem a failure.

This persistence of the status quo is well illustrated by the history of endurance racing—which was initially the province of dedicated riders and dealers rather than factory-backed stars. Once the early problems of reliability were solved, the quest for enhanced all-round performance, free from conservative influences, led to an upsurge of technical innovation. And though some designs were a bit harebrained, others had appreciable merit.

But then the big factories (notably Honda) recognized the publicity value of the championship and brought the full weight of their resources to bear on it—successfully of course. An unfortunate consquence was a stifling of innovation: the manufacturers succeeded with conventional machines and so most other con-

Typical of the innovative design common in endurance racing before the big manufacturers took an interest was the Mead and Tomkinson 'Nessie', with Difazio-type hub-centre steering, underslung fuel tank and unorthodox rear end

Above Trailing-link front suspension on Eric Offen-stadt's endurance-racing Yamaha. The cast-magnesium spindle clamps may be slid along the fork arms for trail adjustment (Robinson)

Honda RCB endurance racer—a conventional layout that dominated the scene for several years once the manufacturers recognized the publicity value of the championship (Robinson)

testants reverted to that type. Perhaps another reason for slow fundamental change is that ultimate roadholding and hence cornering speed may be little affected by such a change.

In the car world, cornering speeds have increased much over the past few decades. Tyres played a large part in this but improvements in suspension geometry were necessary to allow these tyres to do their job properly.

On a bike, so long as the chassis is rigid enough to maintain wheel alignment and the springing and damping system is able to hold

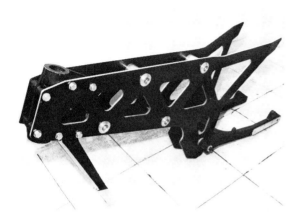

Tony Dawson's bolted-up light-alloy plate frame—
a total manufacturing concept aimed at inexpensive and rapid production of complete machines

the wheels on the ground, then variations in
suspension methods and/or geometry will
have only minor effects on braking and cornering power. Hence racing lap times depend
more on engine, machine weight and tyres
than on revolutionary suspension systems. The
benefit, if indeed there is any, comes in the
stability and feel or handling of the machine.

Before starting any design we must obviously
give careful thought to its overall features—
e.g. size, weight, range, comfort, luggage
space, cost and so on. Then, since several of
the requirements involve conflict, we must
decide how best to compromise. We have
already discussed the design of the various
components; let us now analyse their strength
and stamina.

Strength analysis

With the advent of small, inexpensive com-
puters, the *accurate* analysis of strength and
stiffness in our structures has become

relatively straightforward, given a correct
estimate of the applied loads. Using finite-
element and similar analytical techniques, we
can investigate resonance and vibration
characteristics as well as both overall and
detail stresses and deflections. The very speed
of the computer enables us to try a large number of loading cases and to optimize suspension performance for various types of road
surface.

Our chief problem is to estimate or calculate
the applied loads—and there are several
approaches. We may, for example, choose to
consider 'worst case' conditions and design
for safety under specific maximum loads. This
could mean deciding on the highest speed at
which a head-on collision would not cause
any significant damage; or the speed at which
damage is limited to, say, the front wheel and
fork.

We can check other maximum-load conditions, too, such as hitting a pot-hole or brick
in the road at top speed (perhaps also while
cornering) or landing after a jump of a certain
height with passenger and luggage. Given the
worst-case conditions, these loads can be
reasonably well calculated—but selection of
the conditions is purely arbitrary. If the cases
are too severe, weight will almost certainly be
excessive, whereas underestimation will result
in a fragile structure, intolerant of hard use.

Since metal fatigue is the most common
cause of failure, a study of the high-load conditions gives no indication of the life of a structure in normal service. (Fatigue, as explained
in Chapter 4, is a result of pulsating stress.)

Since conditions of use may vary widely
from machine to machine, it is impossible to
predict the future loading history of any particular frame. The best we can do is to average
out the expected life/load records. Analytical
techniques are of little use here; instead,
exhaustive testing is necessary and that is feasible only for the large manufacturers.

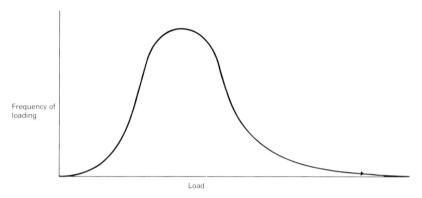

Fig. 8.1 Graph of frame loading against frequency, showing predominance of middle-range loads

A November 1983 Milan Show special—the student-designed Bimota Tesi, with hydraulically operated hub-centre steering. Regardless of its merits and demerits, it would be nice to think that this indicated an increasing acceptance of alternative concepts (Parker)

A bike resembling the new design as closely as possible must be fitted with strain gauges and load transducers, then ridden for a very large mileage under a wide variety of road conditions (or simulated on a rolling road), with records kept of the time history of the loading. Given sufficient testing, we can thus build up a picture of the load cycles of a typical machine of that type over its expected lifetime.

The results can be plotted as shown in figure 8.1.

Note that the middle range of load occurs most frequently, while both low and high values occur much less often. This is to be expected as low loading is produced at low speeds on smooth roads, while high loading is a result of severe conditions—both extremes being encountered for only a relatively small proportion of the machine's life.

Mathematical techniques are available for the treatment of such test data and, together with stress analysis, they enable us to investigate the frame's fatigue behaviour.

Such sophisticated testing methods are beyond the scope of the small chassis maker and the enthusiast building a one-off. Indeed, it is unlikely that even large manufacturers make full use of available techniques. However, good results can usually be achieved by rule-of-thumb and simpler calculations based on previous practice and experience. If the designer himself has the experience, so much the better. If not—and all designers must start somewhere—then a review of existing hardware is a good basis to build on.

After applying appropriate safety factors, allowable stress levels must be decided—which will enable us to determine the sizes of components (frame tubes or whatever).

The need for safety factors arises because of the possible severe consequences of failure; there is always a danger to life and limb if something breaks. Safety factors allow for our inability to predict future loads exactly; for some machines having a much harder life than average; and for any material flaws or inaccuracies in calculation or manufacturing.

The future

Predicting specific design trends is risky but in general there is unlikely to be any radical change in concept so far as mass-produced machines are concerned. As in the past, change will come through small steps on successive models.

There will be advances in materials and manufacturing methods, with greater use of high-strength lightweight plastics and composites. For economic reasons, frames will incorporate fewer tubes and more presswork.

The deplorable upward trend in weight seems to be slowing and it is to be hoped that we can look forward to lighter machines.

As always, we must look to small concerns and talented individuals for serious efforts to develop radical ideas (such as the Elf-sponsored French endurance racers). Unfortunately, however, few brainwaves of that sort will find acceptance in the harsh commercial world.

Index